Excel in Complex Variables

with the Complex Variable Boundary Element Method

About the Authors

Bryce D. Wilkins: Bryce Wilkins graduated from the United States Military Academy in 2017 earning a Bachelor's degree in Mathematical Sciences. He then earned a Master's degree in Mechanical Engineering from the Massachusetts Institute of Technology as a Draper Fellow. As an Army officer, he has attended the Air Defense Artillery Basic Officer Leader Course and the Terminal High Altitude Area Air Defense (THAAD) course at Fort Sill, Oklahoma. In the Army, he was briefly stationed as a Tactical Control Officer for the THAAD battery in Guam. Currently, Bryce is a graduate student at Carnegie Mellon University where he is pursuing a Master's degree in Computational Data Science as part of the inaugural cohort of Army Artificial Intelligence Scholars. Upon graduation, he will work at the Army Artificial Intelligence Task Force in Pittsburgh, Pennsylvania.

Samuel Humphries: Samuel Humphries is currently a Lincoln Laboratory Fellow studying Operations Research at the Massachusetts Institute of Technology. Sam's research interests revolve around leveraging methods in applied math, machine learning, and data analytics to address problems related to climate change and national security. At MIT, Sam is using data-driven techniques to design interventions that increase fuel efficiency in the trucking industry. Sam received his undergraduate degree in Mathematical Sciences from the United States Military Academy. Following his time in graduate school, he will serve as Infantry Officer in the United States Army.

John Lee: John Lee is a Second Lieutenant in the United States Army. He grew up in the Seattle area where he developed an early fascination for mathematics and technology through extracurriculars like math club. Later, he attended the United States Military Academy where he studied mathematics with a focus in applied statistics. John is passionate about using data to provide insights in areas of interest to him. In particular, he is intrigued by the potential of machine learning in the field of political science. In his free time, John enjoys exploring the outdoors, reading about history and politics, and spending time with his family.

Matthew Fauerbach: Matthew Fauerbach is a BS/MS candidate at New York University's Tandon School of Engineering, pursuing a double undergraduate degree in mathematics and computer science with a Master's degree in Cybersecurity. As a CyberCorps SFS scholar, he conducts research through the Center for Cyber Security at NYU Tandon, focusing primarily on post quantum cryptographic primitives and quantum computation at the intersection of applied mathematics and theoretical computer science. Additionally, he is a member of the NYU Courant SIAM chapter and the student-run Offensive Security, Incident Response, and Internet Security lab.

Theodore V. Hromadka II: Dr. Theodore V. Hromadka II is a Distinguished Professor of Mathematics in the Department of Mathematical Sciences at the United States Military Academy. He is also the Principal and Founder of the consulting firm Hromadka & Associates. He holds several professional licenses, including Civil Engineer, Geoscientist, and Geologist. Furthermore, he holds professional certifications as a Hydrologist for both surface water and groundwater. His scientific contributions to the hydrologic, earth, and atmospheric sciences have been widely-published in peer-reviewed scientific literature, including over 30 books and more than 500 scientific papers, book chapters, and government reports. His professional engineering experience includes the supervision and development of over 1,500 engineering studies. He is Professor Emeritus at the California State University, where he served in the Departments of Mathematics, Geological Sciences, and Environmental Studies. Additionally, he serves on the Board of Directors at the Wessex Institute of Technology in the United Kingdom.

Contents

Part I

Introduction to complex variables

Why *Excel* in complex variables?

Bryce D. Wilkins
Theodore V. Hromadka II

The study of complex variables has been a field of endeavor for mathematicians and engineers since the late 1700s. More recently, the use of complex variables as tools for the analysis of potential problems has greatly increased with the advent of modern computational methods. A significant advancement in the use of complex variable theory for the approximation of boundary value problems was introduced in the early 1980s with the publication of the Complex Variable Boundary Element Method (CVBEM) [1, 2, 3, 4]. Since its development, the CVBEM has been extended in a number of directions with useful applications and is well documented in the scientific literature in the form of numerous journal articles and several books [5, 6, 7, 8, 9, 10].

In this book, we seek to achieve two ends. The first is to provide a new take on introductory complex variables by approaching this study using the familiar program Microsoft ®*Excel* (hereafter referred to as "*Excel*" for brevity). The motivation for using *Excel* is that it eliminates the "black box" nature of more advanced computer software and programming languages. Therefore, by implementing the given programs in *Excel*, the reader has the chance to become more familiar with the underlying mathematics of the complex variable programs. The second goal of this book is to provide a comprehensive guide to implementing CVBEM programs in *Excel*.

This book consists of two parts. In Part I, we cover several topics that one would expect to find in any introductory text on complex variables. These topics include an overview of complex numbers, functions of a complex variable, and the Cauchy integral formula. In particular, attention is given to the study of analytic complex variable functions. This attention is warranted because of the property that the real and imaginary parts of an analytic complex variable function can be used to solve the Laplace partial differential equation (PDE). Laplace's equation is ubiquitous throughout science and engineering as it can be used to model the steady-state conditions of several important transport processes including heat transfer, soil-water flow, electrostatics, and ideal fluid flow, among others.

In Part II, we examine a specialty application of complex variables known as the Complex Variable Boundary Element Method, which is a numerical method used for solving boundary value problems governed by Laplace's equation. However, we note that it is not required to have studied PDEs before, as we will provide the necessary details for the specific applications under consideration. Indeed, we have done our best to make the material presented in this book accessible to any interested reader who has previously studied multivariable calculus.

Each chapter combines a succinct presentation of some complex variable theory with a significant focus on examples and using *Excel* to visualize and understand the theory. In this fashion, the steps taken towards the final computational application are demonstrated with all steps fully transparent to the reader. While the market abounds with highly sophisticated computational software that can easily solve the classes of problems examined in this book with amazing visualization outcomes, our belief is that a more valuable learning experience is obtained by eschewing these products in favor of the more deliberate *Excel* approach. Using *Excel* requires working out these myriad problems. Our philosophy for this book is that by using *Excel*, there can be greater illumination in the subject of complex variables.

In fact, the uniqueness of this book is due to the particular emphasis given to developing complex variable programs in *Excel* and, particularly, the development of a CVBEM program in *Excel*, as accomplished in Part II. For all of the *Excel* programs that are featured in this book, the necessary computer code is presented in the text and can also be downloaded from the associated website at www.usmathematics.com.

This book is intended for advanced undergraduate students with an interest in science, engineering, or mathematics. However, this book is not designed to be a stand-alone text on undergraduate-level complex variables. Instead, we discuss only as much of the standard complex variables curriculum as necessary to enable the computational exploration of a selection of important problems with the program *Excel*. We hope you enjoy this exploration into complex variables and the Complex Variable Boundary Element Method aided by the computer program *Excel*!

Chapter 1

Complex numbers

Bryce D. Wilkins
John Lee

Contents

Our study begins with an examination of the arithmetic operations, algebraic properties, and geometric interpretation of complex numbers. To each of these facets of complex numbers, we have given particular emphasis to analogous operations and properties of the real numbers. Additionally, consistent with the theme of this book, each topic discussed is supplemented with an *Excel* implementation.

A basic understanding of arithmetic operations with complex variables is a crucial foundation for further study. With the help of *Excel*, we will explicitly demonstrate the mechanics of these operations. In later chapters, we will use these operations as building blocks to examine more advanced applications of complex variables.

1.1 Arithmetic operations

In this section:

1. The definitions of imaginary and complex numbers.
2. Addition, subtraction, multiplication, and division of complex numbers.
3. *Excel* implementations of arithmetic operations using complex numbers.

We begin this section by establishing some essential definitions for our study.

Definition 1.1.0.1 The **imaginary unit**, denoted i, satisfies $i^2 = -1$.

Definition 1.1.0.2 Let a and b be real numbers. A **complex number**, denoted z, is a number of the form $z = a + bi$. If $b = 0$, then z is a real number. If $a = 0$, then z is said to be **purely imaginary** number.

Definition 1.1.0.3 Given a complex number, $z = a + bi$, we define the **real part of *z*** as follows:

$$\mathrm{Re}\left(z\right) = a$$

Likewise, the **imaginary part of *z*** is defined as follows:

$$\mathrm{Im}\left(z\right) = b$$

Example 1.1.0.1

We will classify the following numbers as either real, purely imaginary, or complex.

(a) π: this number is real because it does not have a nonzero imaginary part.

(b) $3 + \sqrt{2}i$: this number is complex because both the real and imaginary parts are nonzero.

(c) $-5i$: this number is purely imaginary because it does not have a nonzero real part.

Example 1.1.0.2

Recall the numbers considered in Example 1.1.0.1. In this example, we will identify the real and imaginary parts of each of those numbers.

Number	Real Part	Imaginary Part
π	$\text{Re}(\pi) = \pi$	$\text{Im}(\pi) = 0$
$3 + \sqrt{2}i$	$\text{Re}\left(3 + \sqrt{2}i\right) = 3$	$\text{Im}\left(3 + \sqrt{2}i\right) = \sqrt{2}$
$-5i$	$\text{Re}(-5i) = 0$	$\text{Im}(-5i) = -5$

With these few definitions, we are now able to demonstrate how complex numbers work under arithmetic operations. The reader will recognize that many of the arithmetic operations on complex numbers are treated in the same fashion as they are on real numbers.

1.1.1 Addition

Definition 1.1.1.1 Given two complex numbers, $a + bi$ and $c + di$, we define **addition** as follows:

$$(a + bi) + (c + di) = (a + c) + i(b + d)$$

Note 1.1.1.1

Observe that when $b = 0$ and $d = 0$ in Definition 1.1.1.1, the definition of complex addition reduces to the familiar case of addition of two real numbers.

Example 1.1.1.1

$$(1 + 3i) + (2 + 4i) = (1 + 2) + (3 + 4)i = 3 + 7i \tag{1.1}$$

***Excel* Implementation 1.1.1.1**

Cell	A1	B1	C1
Description	First complex number	Second complex number	Sum
Code	=COMPLEX(1,3)	=COMPLEX(2,4)	=IMSUM(A1,B1)
Output	$1 + 3i$	$2 + 4i$	$3 + 7i$

Table 1.1: *Excel* implementation of the problem shown in Example 1.1.1.1.

Notice that the addition of two complex numbers involves adding the real and imaginary parts of each number separately. That is,

$$\begin{aligned} \mathrm{Re}\left[(a + ib) + (c + id)\right] &= \mathrm{Re}\,(a + ib) + \mathrm{Re}\,(c + id) = a + c \\ \mathrm{Im}\left[(a + ib) + (c + id)\right] &= \mathrm{Im}\,(a + ib) + \mathrm{Im}\,(c + id) = b + d. \end{aligned} \tag{1.2}$$

Example 1.1.1.2

We can compute the real and imaginary parts of the sum considered in Example 1.1.1.1 as follows:

$$\begin{aligned} \mathrm{Re}\left[(1 + 3i) + (2 + 4i)\right] &= \mathrm{Re}\,(1 + 3i) + \mathrm{Re}\,(2 + 4i) = 1 + 2 = 3 \\ \mathrm{Im}\left[(1 + 3i) + (2 + 4i)\right] &= \mathrm{Im}\,(1 + 3i) + \mathrm{Im}\,(2 + 4i) = 3 + 4 = 7. \end{aligned} \tag{1.3}$$

***Excel* Implementation 1.1.1.2**

Cell	A1	B1	C1
Description	Real part of first number	Imaginary part of first number	Real part of second number
Code	=1	=3	=2
Output	1	3	2

Cell	D1	E1	F1
Description	Imaginary part of second number	Real part of sum	Imaginary part of sum
Code	=4	=A1+C1	=B1+D1
Output	4	3	7

Table 1.2: *Excel* implementation of Example 1.1.1.2 showing complex addition by real and imaginary parts.

1.1.2 Subtraction

Definition 1.1.2.1 Given two complex numbers, $a + ib$ and $c + id$, we define **subtraction** as follows:

$$(a + ib) - (c + id) = (a - c) + i(b - d) \tag{1.4}$$

Note 1.1.2.1

Observe that when $b = 0$ and $d = 0$ in Definition 1.1.2.1, the definition of complex subtraction reduces to the familiar case of subtraction of two real numbers.

Example 1.1.2.1

$$(1 + 3i) - (2 + 4i) = (1 - 2) + (3 - 4)i = -1 - 1i \tag{1.5}$$

Excel Implementation 1.1.2.1

Cell	A1	B1	C1
Description	First complex number	Second complex number	Difference
Code	=COMPLEX(1,3)	=COMPLEX(2,4)	=IMSUB(A1,B1)
Output	$1+3i$	$2+4i$	$-1-1i$

Table 1.3: *Excel* implementation of the problem shown in Example 1.1.2.1.

As with complex addition, the subtraction of two complex numbers involves subtracting the real and imaginary parts of each number separately. That is,

$$\begin{aligned} \mathrm{Re}\left[(a+ib)-(c+id)\right] &= \mathrm{Re}\,(a+ib)-\mathrm{Re}\,(c+id) = a-c \\ \mathrm{Im}\left[(a+ib)-(c+id)\right] &= \mathrm{Im}\,(a+ib)-\mathrm{Im}\,(c+id) = b-d \end{aligned} \tag{1.6}$$

Example 1.1.2.2

We can compute the real and imaginary parts of the sum considered in Example 1.1.1.1 as follows:

$$\begin{aligned} \mathrm{Re}\left[(1+3i)-(2+4i)\right] &= \mathrm{Re}\,(1+3i)-\mathrm{Re}\,(2+4i) = 1-2 = -1 \\ \mathrm{Im}\left[(1+3i)-(2+4i)\right] &= \mathrm{Im}\,(1+3i)-\mathrm{Im}\,(2+4i) = 3-4 = -1. \end{aligned} \tag{1.7}$$

Excel Implementation 1.1.2.2

Cell	A1	B1	C1
Description	Real part of first number	Imaginary part of first number	Real part of second number
Code	=1	=3	=2
Output	1	3	2

Cell	D1	E1	F1
Description	Imaginary part of second number	Real part of sum	Imaginary part of sum
Code	=4	=SUB(A1,C1)	=SUB(B1,D1)
Output	4	−1	−1

Table 1.4: *Excel* implementation of Example 1.1.2.2 showing complex subtraction by real and imaginary parts.

1.1.3 Multiplication

Let a, b, c, and d denote real numbers. The product $(a+b)(c+d)$ is computed as follows:

$$(a+b)(c+d) = ac + ad + bc + bd \tag{1.8}$$

This procedure can be extended to multiplication with complex numbers.

Definition 1.1.3.1 Let a, b, c, and d denote real numbers. Then, the **product** of $(a+ib)$ and $(c+id)$, is computed by

$$\begin{aligned}(a+bi)(c+di) &= ac + i(ad) + i(bc) + i^2 bd \\ &= ac - bd + i(ad+bc)\end{aligned} \tag{1.9}$$

Note 1.1.3.1

Observe that when $b=0$ and $d=0$ in Definition 1.1.3.1, the definition of complex multiplication reduces to the familiar case of multiplication of two real numbers.

The real and imaginary parts of the product $(a+ib)(c+id)$ are given by, respectively,

$$\begin{aligned}\mathrm{Re}\left[(a+ib)(c+id)\right] &= \mathrm{Re}\,(a+ib)\,\mathrm{Re}\,(c+id) - \mathrm{Im}\,(a+ib)\,\mathrm{Im}\,(c+id) \\ &= ac - bd \\ \mathrm{Im}\left[(a+ib)(c+id)\right] &= \mathrm{Im}\,(a+ib)\,\mathrm{Re}\,(c+id) + \mathrm{Re}\,(a+ib)\,\mathrm{Im}\,(c+id) \\ &= bc + ad\end{aligned} \tag{1.10}$$

Example 1.1.3.1

$$(1+3i)(2+4i) = 2+4i+6i-12 = -10+10i \tag{1.11}$$

***Excel* Implementation 1.1.3.1**

Cell	A1	B1	C1
Description	First complex number	Second complex number	Product
Code	=COMPLEX(1,3)	=COMPLEX(2,4)	=IMPRODUCT(A1,B1)
Output	$1+3i$	$2+4i$	$-10-10i$

Table 1.5: *Excel* implementation of the problem shown in Example 1.1.3.1.

***Excel* Implementation 1.1.3.2**

Cell	A1	B1	C1	D1	E1
Description	First complex number	Second complex number	Product part	Real part	Imaginary part
Code	=COMPLEX(1,3)	=COMPLEX(2,4)	=IMPRODUCT(A1,B1)	=IMREAL(C1)	=IMAGINARY(C1)
Output	$1+3i$	$2+4i$	$-10-10i$	-10	-10

Table 1.6: *Excel* program to compute the real and imaginary parts of the product of two complex numbers.

1.1.4 Division

Definition 1.1.4.1 Let $z = a + bi$ be a complex number. The **complex conjugate** of z, denoted $\bar{z}$, is defined to be:

$$\bar{z} = a - bi$$

Note 1.1.4.1

Let $z = a + bi$ be a complex number. Taking the complex conjugate of z twice gives us the original complex number again such that:

$$\bar{\bar{z}} = z$$

Note 1.1.4.2

Let $z = a + bi$ be a complex number. Then, $z = \bar{z}$ if and only if $b = 0$.

Complex conjugates in *Excel* require a single command, IMCONJUGATE, which takes one complex number as input and provides the conjugate in response, shown below in Table 1.7.

***Excel* Implementation 1.1.4.1**

Cell	A1	B1
Code	=COMPLEX(1,1)	=IMCONJUGATE(A1)
Output	$1 + 1i$	$1 - 1i$

Table 1.7: Complex conjugates.

Dividing complex numbers is the first noteworthy departure from the familiar arithmetic operations on real numbers. To divide complex numbers, multiply the numerator and denominator by the complex conjugate of the denominator.

Definition 1.1.4.2 Let a, b, c, and d denote real numbers. Then, the **quotient** of $(a + ib)$ and $(c + id) \neq 0$ is computed by:

$$\frac{a+ib}{c+id} = \frac{a+ib}{c+id} \cdot \frac{c-id}{c-id} = \frac{ac+bd+i(bc-ad)}{c^2+d^2} \tag{1.12}$$

Note 1.1.4.3

Suppose $c \neq 0$. Then, when $b = 0$ and $d = 0$ in Definition 1.1.4.2, the definition of complex division reduces to the familiar case of division of two real numbers.

$$\frac{a+0i}{c+0i} = \frac{ac+0\cdot 0+i(0\cdot c-a\cdot 0)}{c^2+0^2} = \frac{ac}{c^2} = \frac{a}{c}, \quad c \neq 0. \tag{1.13}$$

Example 1.1.4.1

$$\begin{aligned} \frac{1+3i}{2+4i} &= \frac{1+3i}{2+4i}\cdot\frac{2-4i}{2-4i} \\ &= \frac{2+12+i(6-4)}{2^2+4^2} \\ &= \frac{14+2i}{20} \\ &= \frac{7}{10}+\frac{1}{10}i \end{aligned} \tag{1.14}$$

***Excel* Implementation 1.1.4.2**

Cell	A1	B1	C1
Code	=COMPLEX(1,3)	=COMPLEX(2,4)	=IMDIV(A1,B1)
Output	$1+3i$	$2+4i$	$0.7+0.1i$

Table 1.8: *Excel* implementation of Example 1.1.4.1 showing the division of two complex numbers.

One more helpful note to keep in mind for dividing complex numbers in *Excel* is the syntax for setting up an *if* statement to prevent the user from attempting to divide by zero. This program would be as follows:

Cell	A1	B1	C1
Code	=COMPLEX(1,3)	=COMPLEX(2,4)	=IF(B1=0, "Error: Division by 0", IMDIV(A1,B1))
Output	$1+3i$	$2+4i$	$0.7+0.1i$

Table 1.9: A modified *Excel* implementation of Example 1.1.4.1 that displays an error if the user attempts to divide by zero.

1.2 Algebraic properties

In this section:

1. Additive and multiplicative inverses.

2. Modulus.

Having now defined and explored the fundamental arithmetic operations within the context of complex numbers, it is now necessary to examine more detailed algebraic properties of the complex numbers. Understanding these properties, such as inverses, complex conjugates, and modulus, will help us build the necessary foundation for more advanced applications and computation with complex variables. In later chapters, we will use the material presented in this section to develop useful *Excel* programs.

1.2.1 Additive and multiplicative inverses

The field $\mathbb{C}$ is defined with the operations addition and multiplication. This field has an additive identity, $0 + 0i$, which satisfies the following:

$$z + (0 + 0i) = z, \quad \forall z \in \mathbb{C}. \tag{1.15}$$

The field $\mathbb{C}$ also has a multiplicative identity, $1 + 0i$, which satisfies the following:

$$z \cdot (1 + 0i) = z, \quad \forall z \in \mathbb{C}. \tag{1.16}$$

From the addition and multiplication operations, we can define the familiar notions of subtraction and division of complex numbers. For $z = a + bi$, we define $-z = -(a + bi) = -a - bi$. The numbers z and $-z$ are related by

$$z + -(z) = a + bi + -(a + bi) = a + bi - a - bi = 0. \tag{1.17}$$

Thus, $z + (-z) = z - z = 0$, and we say that $-z$ is the additive inverse of z. An additive inverse exists for all $z \in \mathbb{C}$, and can be computed as $-z$. That is, the additive inverse of a complex number is the negative of itself. Furthermore, to subtract two complex numbers, you add the negative of the second number to the first number: $z_1 - z_2 = z_1 + (-z_2)$.

The notion of an additive inverse for a complex number is very similar to the notion of an additive inverse for a real number. On the contrary, some of the mechanics for multiplicative inverses are quite different in the context of complex numbers than in the context of their real counterparts. Observing the unique method shown in Definition (1.1.4.2) for complex division hints at this disparity. For $z \neq 0 \in \mathbb{C}$, the multiplicative inverse is denoted z^{-1} and satisfies the following relation:

$$z \cdot z^{-1} = 1, \qquad z \neq 0. \tag{1.18}$$

Note 1.2.1.1

The notation z^{-1} does not imply that the multiplicative inverse of a number is necessarily its reciprocal. While this is true for real numbers, in the complex space this is not the case.

Example 1.2.1.1

Let $z \neq 0 \in \mathbb{C}$. We can write $z = a + bi$ for some $a, b \in \mathbb{R}$. Our goal is to derive an expression for z^{-1}. Since we do not currently know the value of z^{-1}, we will write it as $z^{-1} = c + di$. Now, the goal is to determine the values of $c, d \in \mathbb{R}$. We will use the relation in Equation (1.18) to do this. Namely,

$$1 = z \cdot z^{-1} = (a + bi)(c + di) = ac - bd + i(ad + bc). \tag{1.19}$$

Identifying the real and imaginary parts of Equation (1.19), we conclude

$$ac - bd = 1 \qquad \text{and} \qquad ad + bc = 0. \tag{1.20}$$

Recall, a and b are known quantities. Therefore, Equation (1.20) constitutes a system of two equations in two unknowns. Hence, we can derive expressions for c and d in terms of the known quantities a and b. Observe,

$$\begin{bmatrix} a & -b \\ b & a \end{bmatrix} \begin{bmatrix} c \\ d \end{bmatrix} = \begin{bmatrix} 1 \\ 0 \end{bmatrix}. \tag{1.21}$$

Equation (1.21) has a unique solution as long as $\det \begin{bmatrix} a & -b \\ b & a \end{bmatrix} \neq 0$. Recall the determinant for a 2×2 matrix is computed as follows:

$$\det \begin{bmatrix} a & -b \\ b & a \end{bmatrix} = a^2 - (-b^2) = a^2 + b^2. \tag{1.22}$$

Since a and b are real numbers, $a^2 \geq 0$ and $b^2 \geq 0$. In fact, $a^2 = 0$ and $b^2 = 0$ if and only if $a = b = 0$. However, since we assumed $z \neq 0$, at least a or b is nonzero. Thus, we have $a^2 + b^2 > 0$. So, $\det \begin{bmatrix} a & -b \\ b & a \end{bmatrix} \neq 0$, and we conclude that the matrix equation in Equation (1.21) has a unique solution, which is given below:

$$\begin{aligned} c &= \frac{a}{a^2 + b^2} \\ d &= \frac{-b}{a^2 + b^2} \end{aligned} \tag{1.23}$$

It follows that $z^{-1} = \frac{a}{a^2+b^2} - \left(\frac{b}{a^2+b^2}\right) i$.

Thus, we can quickly show a rigorous definition for division of complex numbers: multiplication by the multiplicative inverse of the second number. This formulation is where the process for multiplying by the complex conjugate of the denominator comes from.

1.2.2 Modulus

Definition 1.2.2.1 The **absolute value** or **modulus** of a complex number is defined as follows:

$$|z| = |a + ib| = \sqrt{a^2 + b^2} \tag{1.24}$$

Note 1.2.2.1

The modulus of a real number is equal to its absolute value, as demonstrated below:

$$\sqrt{a^2} = \sqrt{a^2 + (0)^2} = |a + 0i| = |a|. \tag{1.25}$$

The modulus has a number of important properties. For one, the modulus of a complex number is always non-negative. Also notable is that from the modulus, we can confirm the triangle inequality holds for complex numbers just as it does for real numbers.

Theorem 1.2.2.2 (The Triangle Inequality) *Let* $z_1, z_2 \in \mathbb{C}$*. Then, these numbers satisfy the following inequality:*

$$|z_1 + z_2| \leq |z_1| + |z_2|. \tag{1.26}$$

Note 1.2.2.2

At the end of Example 1.2.1.1, we derived the following expression for the inverse of a complex number $z = a + bi$:

$$z^{-1} = \frac{a}{a^2 + b^2} - \left(\frac{b}{a^2 + b^2}\right)i. \tag{1.27}$$

We can now update this definition using the definition of modulus in Definition 1.2.2.1. Thus,

$$z^{-1} = \frac{a}{|z|^2} - \frac{b}{|z|^2}i = \frac{a - bi}{|z|^2} = \frac{\overline{z}}{|z|^2}. \tag{1.28}$$

1.2.3 Examples

The following are some basic examples of the aforementioned algebraic properties with $z = 1 + i$:

$$(-z) = -(1 + i) = -1 - i \tag{1.29}$$

$$z^{-1} = \frac{1}{1^2 + 1^2} + \frac{-1}{1^2 + 12}i = \frac{1}{2} - \frac{1}{2}i \tag{1.30}$$

$$\overline{z} = \overline{1 + i} = 1 - i \tag{1.31}$$

$$|z| = \sqrt{1^2 + 1^2} = \sqrt{2} \tag{1.32}$$

1.2.4 *Excel* computation

With the help of *Excel*, we can render the calculation of these algebraic properties as trivial. In fact, two of these properties can be derived from a single command.

Additive and multiplicative inverses

As previously mentioned, the additive inverse of a complex number is the negation of that complex number. In particular, the additive inverse of $z = a + bi$ is $-z = -a - bi$. This can be computed in *Excel* by considering the real and imaginary parts of z separately, negating each part, then putting them back together, as shown in Table 1.10.

Excel Implementation 1.2.4.1

Cell	A1	B1	C1
Code	=COMPLEX(1,1)	=IMREAL(A1)	IMAGINARY(A1)
Output	$1 + 1i$	1	1
Cell	D1	E1	F1
Code	-B1	-C1	COMPLEX(D1,E1)
Output	-1	-1	$-1 - i$

Table 1.10: Additive inverses.

In the case of multiplicative inverses, we follow a similar process, only this time we substitute our formula $z^{-1} = \frac{a}{a^2+b^2} - (\frac{b}{a^2+b^2})i$ rather than just negating both parts. We demonstrate this process in Table 1.11.

Excel Implementation 1.2.4.2

Cell	A1	B1	C1
Code	=COMPLEX(1,1)	=IMREAL(A1)	IMAGINARY(A1)
Output	$1 + 1i$	1	1
Cell	D1	E1	F1
Code	=B2/ (POWER(B1,2) +POWER(C1,2))	=-C2/ (POWER(B1,2)+ POWER(C1,2))	=COMPLEX(D1, E1)
Output	0.5	-0.5	$0.5 - 0.5i$

Table 1.11: Multiplicative inverses.

Note 1.2.4.1

The POWER( ) command is a built-in function of *Excel* that raises a real number by a specified exponent. IMPOWER( ) is the equivalent command in *Excel* for complex arguments.

Modulus

Computing the modulus of a complex number in *Excel* can be done using the IMABS( ) function as demonstrated below.

***Excel* Implementation 1.2.4.3**

Cell	A1	B1
Code	=COMPLEX(1,1)	=IMABS(A1)
Output	$1 + 1i$	1.414214

Table 1.12: Modulus.

The result is an approximation of $\sqrt{2}$, as desired.

1.3 Geometric interpretation

In this section:

1. Euler's formula.
2. Principal value of a complex number.
3. n^{th} roots of unity.

Up to this point, we have discussed complex numbers exclusively in the form $z = a + bi$. In this section, we will delve into another approach for representing complex numbers, and we will explore some beneficial properties and consequences of that approach.

1.3.1 Graphical interpretation and the polar form

Since a complex number $z = a + bi$ consists of real and imaginary parts, we can think of a complex number as representing a point in a plane with a real axis and an imaginary axis. Using this coordinate system, z is represented as the ordered pair (a, b).

Note that in this depiction, the x-axis represents the real part and the y-axis represents the imaginary part, such that $z = a + bi$ coincides with the point (a, b).

Now consider if we interpret z in polar coordinates rather than in Cartesian coordinates. In this case, the coordinate (a, b) would instead be (r, θ), where r is the distance (modulus) from the origin to the point and θ is the angle in radians between the positive x axis and a line connecting the origin to the radius.

Note 1.3.1.1

Observe that θ can take infinitely many values since each complete rotation about the origin represents a new value of the angle. That is, (r, θ) and $(r, \theta + 2n\pi)$ represent the same point for any integer n. Therefore, we say that θ is unique up to an integer multiple of 2π.

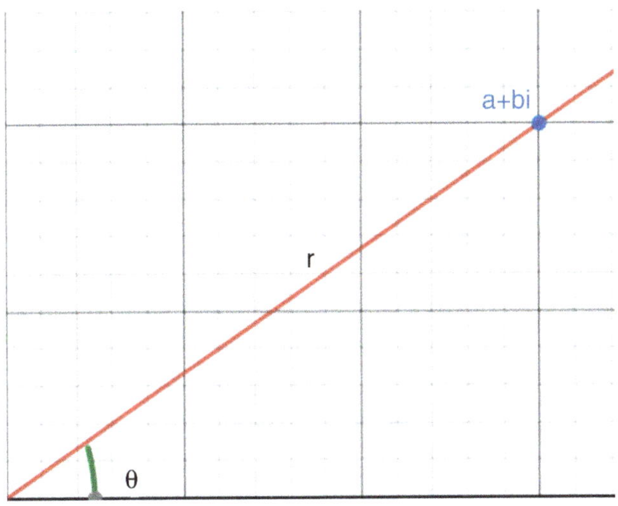

Figure 1.1: The complex number $a + ib$ can be represented in rectangular coordinates in the complex plane by the point (a, b) or can be represented in polar coordinates by (r, θ), where $r = \sqrt{a^2 + b^2}$ and $\theta = \tan^{-1}\left(\frac{b}{a}\right)$. If $a = 0$, then θ is determined by the sign of b. Specifically, if $b > 0$, then $\theta = \pi/2$. If $b < 0$, then $\theta = 3\pi/2$. If $b = 0$, then there is no unique value for θ.

The conversion equations for cartesian to polar form are:

$$x = r\cos(\theta) \quad \text{and} \quad y = r\sin(\theta). \tag{1.33}$$

So if we substitute x and y for a and b in the complex number $z = a + bi$, we obtain the *polar form* of complex numbers:

$$z = r\left(\cos(\theta) + i\sin(\theta)\right). \tag{1.34}$$

While *Excel* does not have a method for converting between forms of complex numbers directly, doing so is easy enough using the IMARGUMENT function and the IMABS function.

***Excel* Implementation 1.3.1.1**

Cell	A1	B1
Code	=COMPLEX(1,1)	=IMARGUMENT(A1)
Output	$1 + i$	0.7853
Cell	C1	D1
Code	=IMABS(A1)	=COMPLEX(COS(B1),SIN(B1))
Output	1.4142	$0.7071 + 0.7071i$

Table 1.13: Polar form.

The last step is to multiply the result in cell D1 by the modulus computed in cell C1.

1.3.2 The argument and the principal form

Definition 1.3.2.1 Let $z = a + bi$. The **argument** of z, denoted $\arg(z)$, is defined as the angle that the line segment connecting $(0, 0)$ and (a, b) makes with the positive real axis.

Note 1.3.2.1

Alternatively, the argument of the complex number $z = a + bi$ can be defined using the polar expression of z:

$$z = r(\cos\theta + i\sin\theta) = re^{i\theta}, \quad r = \sqrt{a^2 + b^2}. \tag{1.35}$$

In this case, we have $\arg(z) = \theta$.

Example 1.3.2.1

Compute the argument of $z = 1 - 1i$.

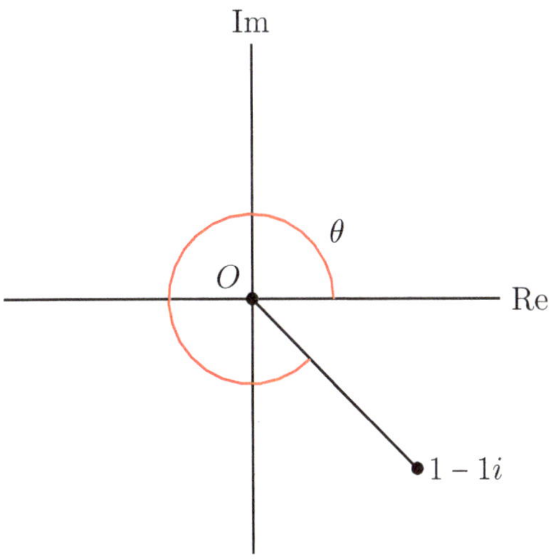

Figure 1.2: The desired angle is denoted θ.

This calculation can be simplified by recognizing that quadrants I, II, and III each contribute $\pi/2$ radians to the total angle. Therefore, it only remains to calculate the portion of θ that is contained in the fourth quadrant.

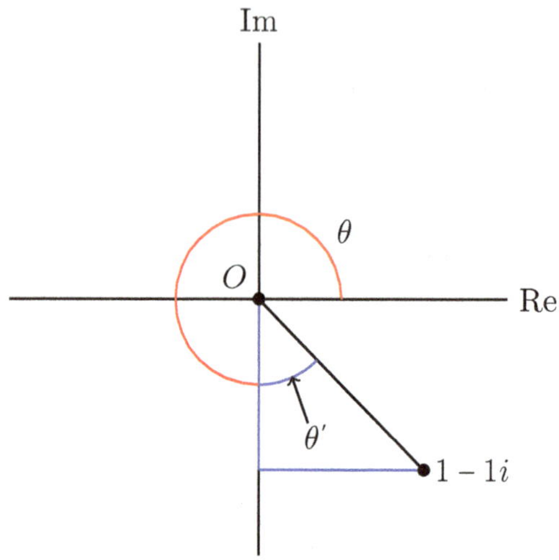

Figure 1.3: The desired angle can be computed as $\theta = \frac{3\pi}{2} + \theta'$, where θ' is the angle denoted by the blue arc.

Using trigonometry, we obtain $\theta' = \tan^{-1}\left(\frac{1}{1}\right) = \frac{\pi}{4}$. Therefore,

$$\arg z = \theta = \frac{3\pi}{2} + \frac{\pi}{4} = \frac{7\pi}{4}.$$

Consider Equation (1.35), and recall that the cosine and sine functions are 2π periodic such that:

$$\begin{aligned} \cos(\theta) &= \cos(\theta + 2\pi) && \text{for all } \theta \in \mathbb{R} \\ \sin(\theta) &= \sin(\theta + 2\pi) && \text{for all } \theta \in \mathbb{R}. \end{aligned}$$

Therefore, the definition of $\arg(z)$ is ambiguous because if a certain θ satisfies $\arg(z) = \theta$, then $\theta + 2\pi n$ could also be considered the argument of z for any integer n. To deal with this ambiguity, we define the principal value of the complex number z, below.

Definition 1.3.2.2 Let $z = a + bi$. The **principal value** of the complex number argument of z, denoted $\text{Arg}(z)$, is defined as the unique angle in the interval $(-\pi, \pi]$ that the line segment connecting $(0,0)$ and (a,b) makes with the positive real axis.

Note 1.3.2.2

We can determine the principal value of the complex number argument of a particular value of $z = a + bi$ as follows:

$$\operatorname{Arg}(z) = \begin{cases} \tan^{-1}(b/a), & a > 0 \\ \tan^{-1}(b/a) + \pi, & a < 0 \text{ and } b \geq 0 \\ \tan^{-1}(b/a) - \pi, & a < 0 \text{ and } b < 0 \\ \pi/2, & a = 0 \text{ and } b > 0 \\ -\pi/2, & a = 0 \text{ and } b < 0 \\ \text{undefined}, & a = 0 \text{ and } b = 0. \end{cases} \tag{1.36}$$

In some programming languages, the computation of $\operatorname{Arg}(z)$ as described in Equation (1.36) is performed by a function called the two-argument arctangent function, commonly written as atan2. *Excel* has such a function.

Example 1.3.2.2

Compute the principal value of the complex argument of $z = 3 - 2i$.

$$\operatorname{Arg}(z) = \tan^{-1}\left(-\frac{2}{3}\right) = -0.588003.$$

The argument is an important calculation in complex variables, as it is needed for both the polar and exponential expressions. Fortunately, *Excel* has a built in function for calculating the argument of a complex number, IMARGUMENT, demonstrated below in Table 1.14.

***Excel* Implementation 1.3.2.1**

Cell	A1	B1
Code	=COMPLEX(1,1)	=IMARGUMENT(A1)
Output	$1 + i$	0.7853

Table 1.14: Argument.

Note 1.3.2.3

Be aware that some authors define the principal value of the complex number argument of z, denoted Arg (z), as the unique angle in the interval $[0, 2\pi)$ that the line segment connecting $(0, 0)$ and (a, b) makes with the positive real axis. This distinction is often a matter of personal preference and will not be important for the purpose of our study.

1.3.3 Euler's formula and exponential form

Definition 1.3.3.1 Let $x \in \mathbb{R}$. **Euler's formula** relates the complex exponential function to sine and cosine as follows:

$$e^{ix} = \cos(x) + i\sin(x) \tag{1.37}$$

More generally, let $z \in \mathbb{C}$. Euler's formula is also valid for complex arguments. Thus,

$$e^{iz} = \cos(z) + i\sin(z) \tag{1.38}$$

Example 1.3.3.1

Compute $e^{i\pi}$. Using Equation (1.37) with $x = \pi$, we obtain

$$\begin{aligned} e^{i\pi} &= \cos(\pi) + i\sin(\pi) \\ &= -1 + i(0) \\ &= -1. \end{aligned}$$

This is a famous result that is sometimes referred to, specifically, as Euler's identity.

Note 1.3.3.1

Equation (1.37) can be thought of as a way of defining the complex exponential function in terms of the real-variable trigonometric functions sine and cosine. However, we can also use Equation (1.37) to create complex exponential definitions of sine and cosine. First, we consider Equation (1.37) with $-x$ as the argument of the complex exponential function:

$$e^{-ix} = \cos(-x) + i\sin(-x) = \cos(x) - i\sin(x). \tag{1.39}$$

Now, adding Equations (1.37) and (1.39), we obtain:

$$e^{ix} + e^{-ix} = 2\cos(x) \implies \boxed{\cos(x) = \frac{e^{ix} + e^{-ix}}{2}}. \tag{1.40}$$

By subtracting Equations (1.37) and (1.39), we obtain:

$$e^{ix} - e^{-ix} = 2i\sin(x) \implies \boxed{\sin(x) = \frac{e^{ix} - e^{-ix}}{2i}}. \tag{1.41}$$

Further, recall that the hyperbolic cosine function is defined as $\cosh(x) = \cos(ix)$, and the hyperbolic sine function is defined as $\sinh(x) = -i\sin(ix)$. We can obtain exponential definitions of these hyperbolic functions by considering imaginary arguments of Equations (1.40) and (1.41). Thus,

$$\cosh(x) = \cos(ix) = \frac{e^{i(ix)} + e^{-i(ix)}}{2} = \frac{e^{-x} + e^{x}}{2}$$

$$\sinh(x) = -i\sin(ix) = -i\frac{e^{i(ix)} - e^{-i(ix)}}{2i} = -\frac{e^{-x} - e^{x}}{2} = \frac{e^{x} - e^{-x}}{2}.$$

Recall that in Section 1.3.1, we showed that all complex numbers, except for $z = 0$, can be written in polar form:

$$z = r\left[\cos(\theta) + i\sin(\theta)\right].$$

Now, using Euler's formula, it is straightforward to obtain a new form of writing a complex number; namely, the exponential form, which is obtained by substituting Euler's formula into the polar form. Observe,

$$z = r\left[\cos(\theta) + i\sin(\theta)\right] = re^{i\theta}. \tag{1.42}$$

Note 1.3.3.2

As we have mentioned before, due to the 2π-periodicity of cosine and sine, there is not a unique representation of a complex number in polar form. Since the exponential form is an alternative way of writing the polar expression of a complex number, it follows that there is also not a unique representation of a complex number in exponential form. To emphasize this non-uniqueness, the exponential form of a complex number is often written as

$$z = r\left[\cos(\theta) + i\sin(\theta)\right] = re^{i\theta} = re^{i(\theta+2\pi k)}, \quad k \in \mathbb{Z}. \tag{1.43}$$

Example 1.3.3.2

Show that $|z| = r$. We begin by writing z in the exponential form: $z = re^{i\theta}$. Then, it follows that

$$\begin{aligned} |z| &= \left|re^{i\theta}\right| \\ &= |r|\left|e^{i\theta}\right| \\ &= |r|\left|\cos(\theta) + i\sin(\theta)\right| \\ &= \sqrt{r^2}\sqrt{\cos(\theta)^2 + \sin(\theta)^2} \\ &= r. \end{aligned}$$

Thus, we have shown that $|z| = r$.

***Excel* Implementation 1.3.3.1**

Cell	A1	B1
Code	=COMPLEX(1,1)	=IMARGUMENT(A1)
Output	$1 + i$	0.7853

Cell	C1
Code	=IMABS(A1)
Output	1.4142

Table 1.15: Exponential form.

Cell B1 contains the argument of the complex number and cell C1 contains the modulus of the complex number. The exponential form is obtained in *Excel* by computing the product of C1 and the quantity of e raised to the power of i times the value of B1. In *Excel*, this is computed as follows: =IMPRODUCT(C1, IMEXP(IMPRODUCT(COMPLEX(0,1),B1))).

Chapter 2

Functions of a complex variable

Bryce D. Wilkins
Samuel Humphries

Contents

Definition 2.0.0.1 Let $\Omega \subseteq \mathbb{C}$. A **complex variable function on** $\boldsymbol{\Omega}$, denoted $f : \Omega \to \mathbb{C}$ is a rule assigning a $w \in \mathbb{C}$ to each $z \in \Omega$. The rule is written as $f(z) = w$.

Definition 2.0.0.2 Let $\Omega \subseteq \mathbb{C}$, and let $f : \Omega \to \mathbb{C}$ be a complex variable function. The set Ω is known as the **domain of definition** of f, and the set $\{w : w = f(z) \text{ for some } z \in \Omega\}$ is known as the **range** of f.

Note 2.0.0.1

For a complex variable function, the domain of definition of the function and the range of the function are both subsets of $\mathbb{C}$.

Complex functions and, in particular, analytic functions, comprise a significant portion of the complex analysis that will be important for the more advanced applications of complex variables in Part II of this book.

In this chapter, we will explore various properties of complex variable functions. We will also consider the special class of complex variable functions known as analytic functions. Finally, we will examine the Taylor series representations of several important complex variable functions, taking time to consider how *Excel* can be leveraged to better understand each of these elementary functions.

2.1 Visualizing complex variable functions

In this section:

1. Domain coloring.

2. Transformation by $1/z$.

One of the first steps to understanding a particular function is to try to visualize it. For one-dimensional real variable functions, this type of visualization can be crucial and can reveal information to the viewer that may not be readily apparent just from the definition of the function. For example, consider the real variable function $g : \mathbb{R} \to \mathbb{R}$ given by $g(x) = \frac{x^3}{3} + x^2 - 5$, as pictured below.

By visualizing g graphically, we can immediately deduce information about it. For example, g has one root on the interval $[-4, 4]$, and the root is approximately $x \approx 1.77$. Moreover, $g \leq 0$ on the interval $[-4, 1.77]$ and $g > 0$ on the interval $(1.77, 4]$. We can also deduce that on the interval $[-4, 4]$: g has two inflection points, has a minimum of approximately -10, and has a maximum of approximately 30. All of these conclusions are things that we can easily determine just by looking at the graph of g, but we could not as quickly determine them by looking at the definition of g alone.

Now, consider a complex variable function $f : \mathbb{C} \to \mathbb{C}$ given by $f(z) = w$, where $z = x + iy$ and $w = u + iv$. A complication of creating a graphical representation for f is that both z and w are located in the complex plane as opposed to on the real line.

The function g is very easy to visualize using standard graphical techniques because the independent variable x and the value $g(x)$ are both one-dimensional. Such functions can be plotted in a plane, as visualized in Figure 2.1. As a rule of thumb, if the input of a function is n_1-dimensional and the output of a function is n_2-dimensional, then producing a graph of the function requires $n_1 + n_2$ dimensions. In this example, the input of g is one-dimensional and the output of g is one-dimensional, so it takes $1 + 1 = 2$ dimensions to plot.

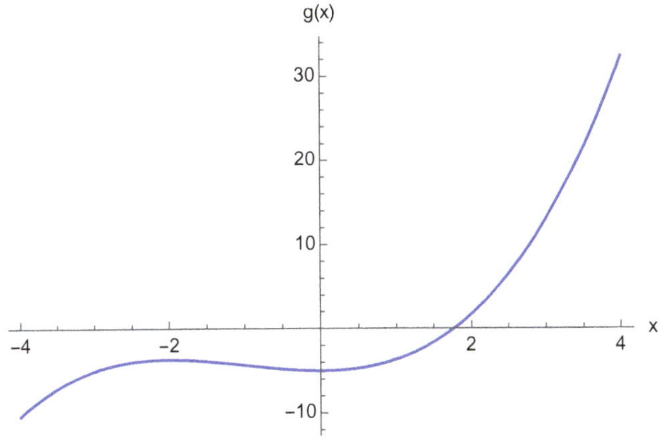

Figure 2.1: A real variable function given by $g(x) = \frac{x^3}{3} + x^2 - 5$.

Now, consider the case of the complex variable function, f, as defined above. In this case, the input of f has two dimensions; namely, x and y. Likewise, the output of f also has two dimensions; namely, u and v. Therefore, using the rule of thumb as just described, we would need $2 + 2 = 4$ dimensions to visualize a complex variable function using the traditional graphical methods.

One creative way to visualize complex variable functions is to use a technique known as domain coloring, as demonstrated in Example 2.1.0.1. Remember, visualizing complex variable functions requires four different dimensions. In the case of domain coloring, two of these dimensions are visualized spatially, one is represented by color, and the last one is represented by the brightness or saturation of the color.

A domain coloring for a given function $f : \mathbb{C} \to \mathbb{C}$ can be interpreted according to the following steps:

1. Identify a point in the complex plane, denoted $z_0 \in \mathbb{C}$.

2. Evaluate the function $f(z_0)$.

3. Determine the color of the value $f(z_0)$ according to a reference coloring of the complex plane.

4. Plot the color of $f(z_0)$ determined in step 3 at the point z_0 in the plane of the domain coloring for f.

We will examine this procedure for several typical complex variable functions in Example 2.1.0.1.

Example 2.1.0.1

The figures below appear in pairs of two. The figure on the left shows a reference coloring of the complex plane. In other words, this is the coloring of the identity function, $f(z) = z$. The figure on the right is the domain coloring for typical examples of some complex variable functions; in particular, we examine a power function, a rational function, and a trigonometric function.

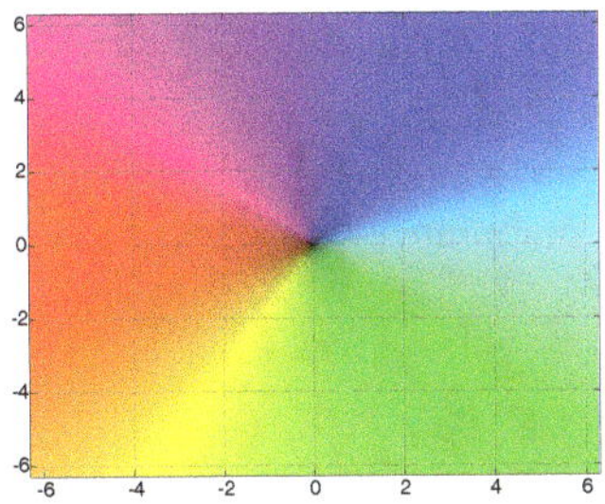

Figure 2.2: Reference coloring of the complex plane.

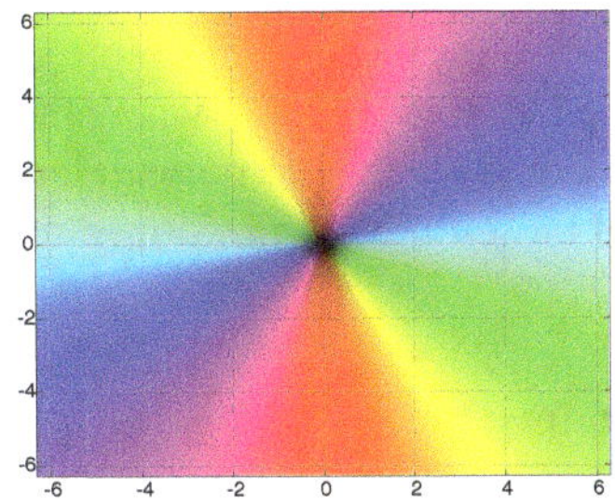

Figure 2.3: Domain coloring for $f : \mathbb{C} \to \mathbb{C}$ given by $f(z) = z^2$.

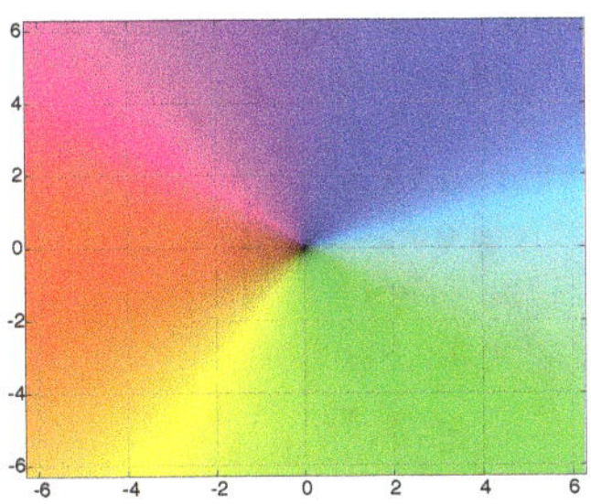

Figure 2.4: Reference coloring of the complex plane.

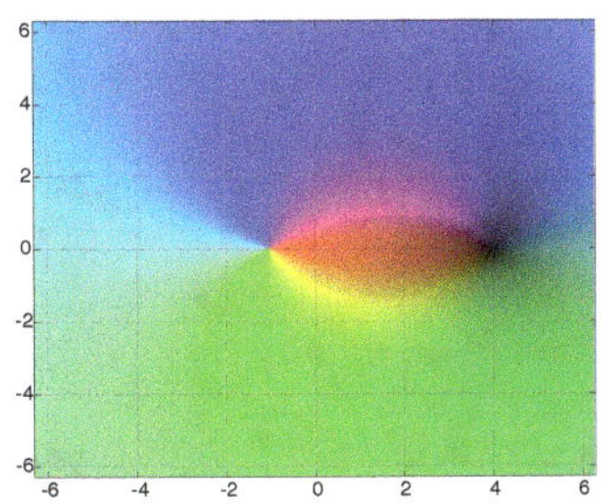

Figure 2.5: Domain coloring for $f : \mathbb{C} \to \mathbb{C}$ given by $f(z) = \frac{z-4}{z+1}$.

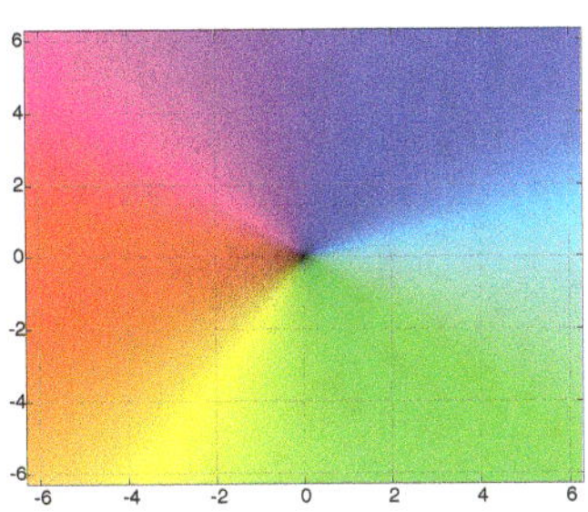

Figure 2.6: Reference coloring of the complex plane.

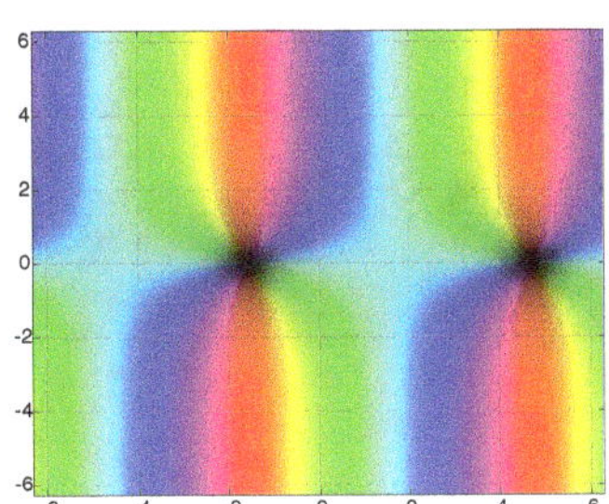

Figure 2.7: Domain coloring for $f : \mathbb{C} \to \mathbb{C}$ given by $f(z) = \sin(z) + 1$.

Another way to visualize complex variable functions is as a mapping from one complex plane (i.e. the input plane) to another complex plane (i.e. the output plane). The next example considers a visualization of the complex variable function $f(z) = \frac{1}{z}$ where the input plane (shown on the left) is colored in concentric rings. The output plane (shown on the right) demonstrates how each of those rings is transformed under the mapping $z \mapsto \frac{1}{z}$.

Example 2.1.0.2

1. This example includes three sets of figures. Each set depicts a different view of the transformation $z \mapsto 1/z$.

2. All of the figures use rings of color to illustrate this transformation. The motivation for this visualization technique is that rings of color are preserved as rings under this particular transformation.

3. Colors that are close to the circle, i.e. $(1-\varepsilon) < |z| \leq 1$ are mapped to just outside of the unit circle under the transformation. Meanwhile, colors that are close to the origin, $z = 0$, are mapped close to infinity (in all directions) under the transformation.

4. In each set of figures, the left figure depicts the unit circle before transformation. The right figure shows the exterior of the unit circle, which is where the unit circle is mapped to under the transformation $z \mapsto 1/z$.

5. The left figure, which depicts the unit circle, is always plotted in the square $[-1, 1] \times [-1, 1]$. However, the right figure is plotted on squares of size 2×2, 4×4, and 6×6 to illustrate that the unit circle is mapped to the set $\mathbb{C} \setminus (|z| \leq 1)$ under this transformation.

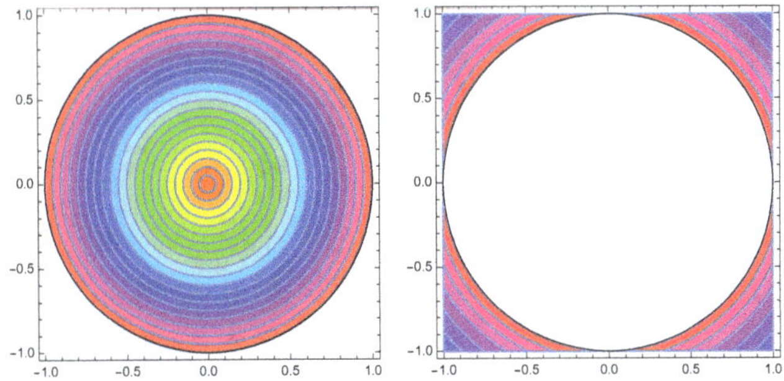

Figure 2.8: Visualizing the transformation $z \mapsto 1/z$. The figure on the left shows rings of the complex plane that are identified by color. The figure on the right shows how those rings are mapped under the transformation $1/z$. The output of the transformation is depicted on the set $[-1,1]\times[-1,1]$.

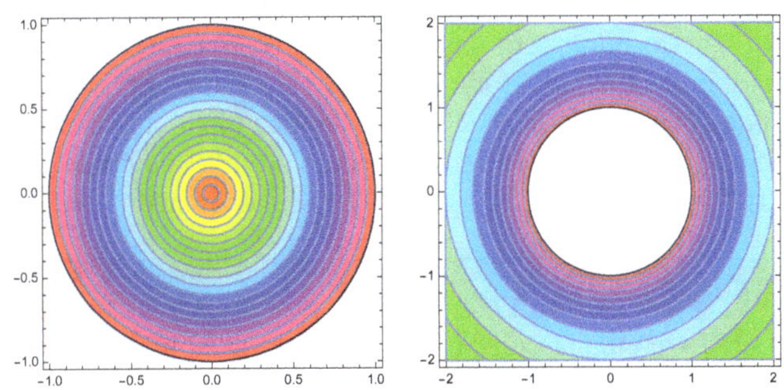

Figure 2.9: Visualizing the transformation $z \mapsto 1/z$. The figure on the left shows rings of the complex plane that are identified by color. The figure on the right shows how those rings are mapped under the transformation $1/z$. The output of the transformation is depicted on the set $[-2,2]\times[-2,2]$.

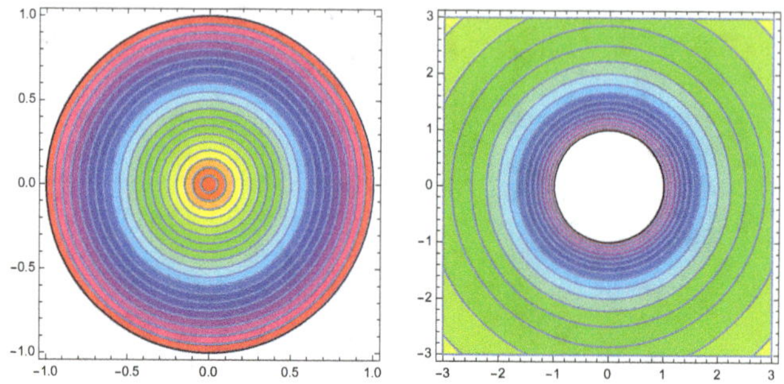

Figure 2.10: Visualizing the transformation $z \mapsto 1/z$. The figure on the left shows rings of the complex plane that are identified by color. The figure on the right shows how those rings are mapped under the transformation $1/z$. The output of the transformation is depicted on the set $[-3, 3] \times [-3, 3]$.

Now, we consider an example that will be helpful later for learning about the CVBEM. For the purpose of this example, it is not important to understand the mechanics of the CVBEM (or even to have any familiarity with it). Rather, it shall suffice to know that the CVBEM is applied to problems with a simple, closed boundary, and the CVBEM invokes the use of "computational nodes," which act as degrees of freedom in the CVBEM modeling process.

Note 2.1.0.1

Disclaimer: To re-emphasize, the graphics that follow in Example 2.1.0.3 are only meant to help demonstrate possible visualization techniques for complex variable functions and are not meant to be instructive about the CVBEM. That will come later!

Example 2.1.0.3

1. This example includes 8 figures. Each figure depicts a different view of the transformation $z \mapsto 1/z$.

2. All of the figures show how a standard geometrical shape is mapped under the transformation $z \mapsto 1/z$.

3. Red is used to depict the shape in the xy plane prior to transformation. Blue is used to depict the shape in the uv plane after the transformation.

4. The shapes before and after transformation are plotted on the same axes to save space and emphasize the difference in scale between the shapes before and after mapping under the transformation $z \mapsto 1/z$.

5. Each figure also shows the transformation of 20 computational nodes located in a circular fashion in the exterior of the original (pre-transformation) problem domain.

6. **Black** is used to denote the computational nodes before transformation. Green is used to denote the computational nodes after transformation.

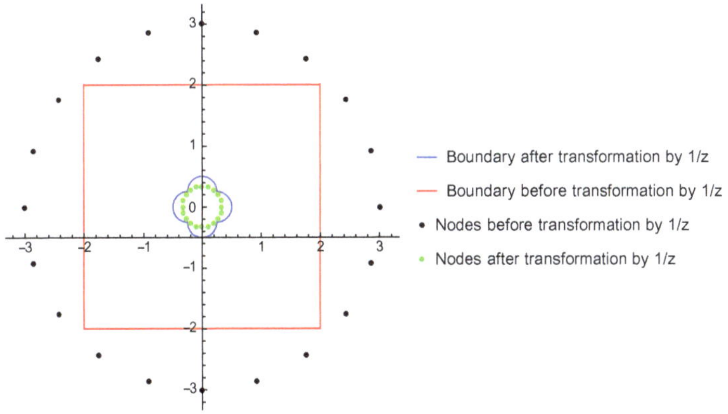

Figure 2.11: Visualizing the transformation $z \mapsto 1/z$. This figure demonstrates how a square is mapped under the transformation. The figure also depicts how a circular arrangement of computational nodes are mapped under the transformation.

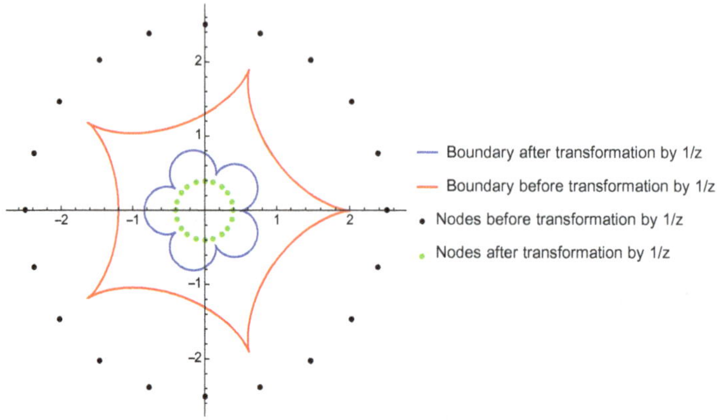

Figure 2.12: Visualizing the transformation $z \mapsto 1/z$. This figure demonstrates how a hypocycloid with five sides is mapped under the transformation. The figure also depicts how a circular arrangement of computational nodes are mapped under the transformation.

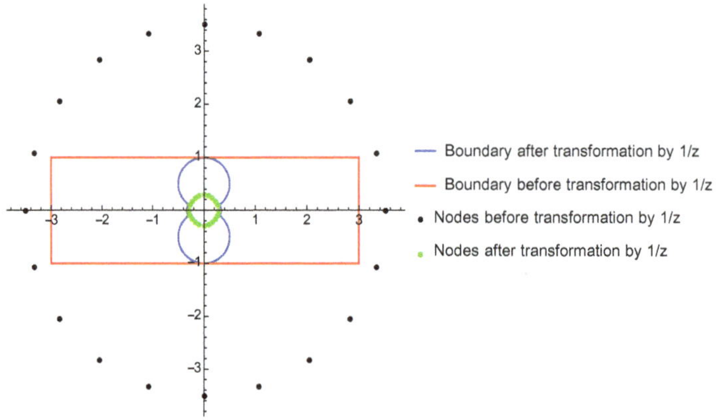

Figure 2.13: Visualizing the transformation $z \mapsto 1/z$. This figure demonstrates how a rectangle is mapped under the transformation. The figure also depicts how a circular arrangement of computational nodes are mapped under the transformation.

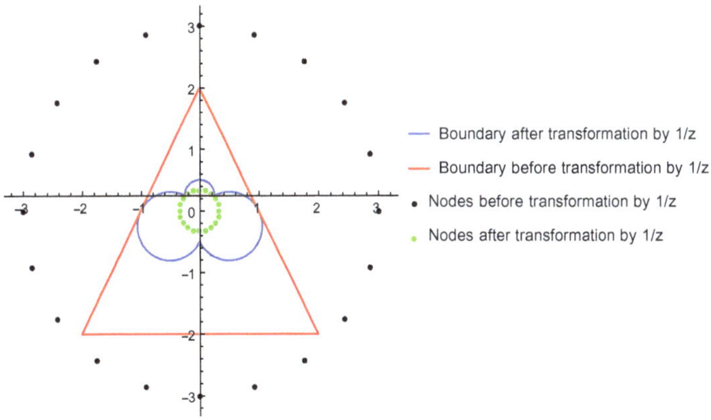

Figure 2.14: Visualizing the transformation $z \mapsto 1/z$. This figure demonstrates how an isosceles triangle is mapped under the transformation. The figure also depicts how a circular arrangement of computational nodes are mapped under the transformation.

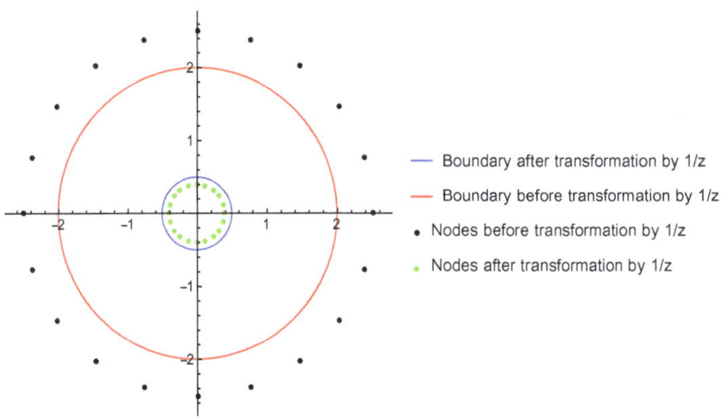

Figure 2.15: Visualizing the transformation $z \mapsto 1/z$. This figure demonstrates how a circle of radius 2 is mapped under the transformation. The figure also depicts how a circular arrangement of computational nodes are mapped under the transformation.

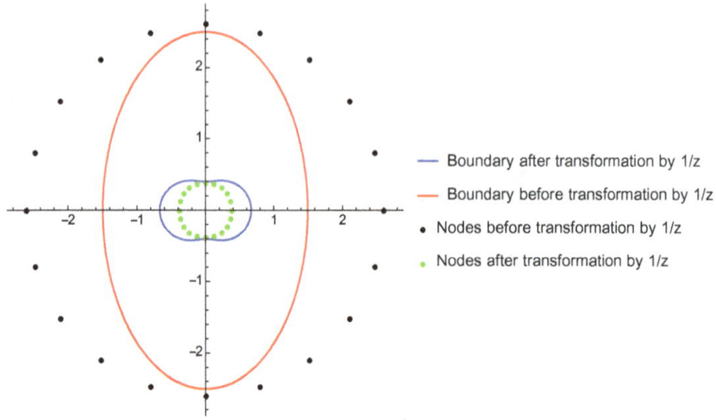

Figure 2.16: Visualizing the transformation $z \mapsto 1/z$. This figure demonstrates how an ellipse is mapped under the transformation. The figure also depicts how a circular arrangement of computational nodes are mapped under the transformation.

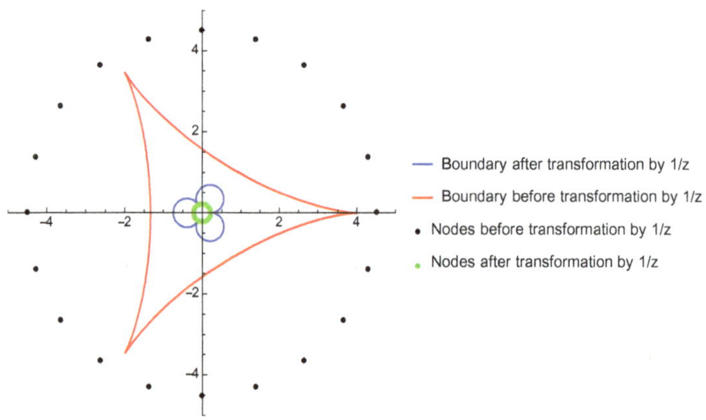

Figure 2.17: Visualizing the transformation $z \mapsto 1/z$. This figure demonstrates how a hypocycloid with three sides is mapped under the transformation. The figure also depicts how a circular arrangement of computational nodes are mapped under the transformation.

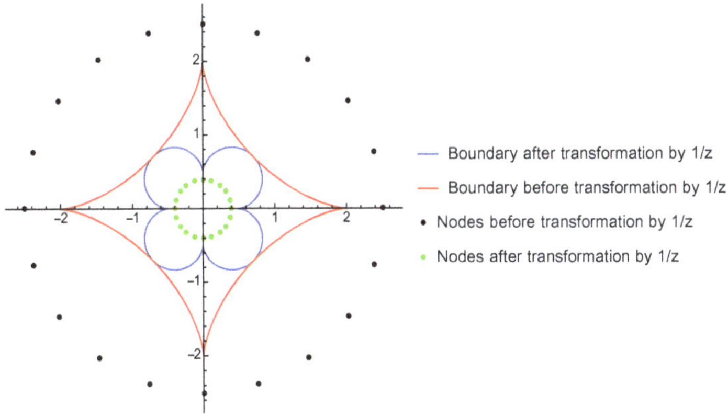

Figure 2.18: Visualizing the transformation $z \mapsto 1/z$. This figure demonstrates how a hypocycloid with four sides is mapped under the transformation. The figure also depicts how a circular arrangement of computational nodes are mapped under the transformation.

2.2 Linear transformations and mappings

In this section:

1. Translation.
2. Rotation.
3. Magnification.

Consider the function f: $D \to C$ given by $f(z) = w$. In order to best visualize both z and w, it is more convenient to draw the two planes separately. From this we get the word mapping, as each z value is displayed, or "mapped", onto a corresponding $f(z)$ or w value, the image of z. As already mentioned, the image of the entire domain D is known as the range of f.

Of note are the different types of mapping, or transformations, that can occur. These types include translation, rotation, and magnification. Each type of transformation has a unique impact on the resulting image of a complex variable.

First, translation is a one-to-one, rigid transformation that displaces z through a vector $T = a + ib$ to a new position, $w = f(z)$. The mathematical representation of a translation can be seen below, where f is the transforming function:

$$w = f(z) = z + T = (x + a) + i(y + b). \tag{2.1}$$

The inverse mapping for a translation is also a one-to-one mapping of the z-plane onto the w-plane. It is displayed mathematically below:

$$z = f^{-1}(w) = w - T = (u - a) + i(v - b). \tag{2.2}$$

Next, a rotation is another one-to-one transformation where z is spun around the origin by an angle α to a new position $w = f(z)$. Mathematically, rotation is expressed as:

$$w = f(z) = ze^{i\alpha} = re^{i\theta}e^{i\alpha} = re^{i(\theta+\alpha)}, \tag{2.3}$$

where $r = |z|$ is the distance from the origin to z.

Using polar coordinates and letting $w = \rho e^{i\phi}$, the inverse mapping becomes:

$$z = f^{-1}(w) = we^{-i\alpha} = \rho e^{i\phi}e^{-i\alpha} = \rho e^{-i(\phi-\alpha)}. \tag{2.4}$$

Yet another one-to-one transformation is magnification, which either enlarges an image or shrinks it. Mathematically, it is written as:

$$w = f(z) = kz = kx + iky, \quad k \in \mathbb{R}. \tag{2.5}$$

If $k < 1$, then the transformation shrinks the distance between points. Conversely, when $k > 1$ the distance between points is increased. The inverse mapping of this transformation is written as:

$$z = f^{-1}(w) = \frac{1}{k}w = \frac{1}{k}u + i\frac{1}{k}v. \tag{2.6}$$

Finally, a linear transformation can be a combination of all three of the above mapping types to create a unique transformation. A linear transformation is a similarity mapping in the sense that an object is mapped to an image similar to itself. It does this by rotating an object through angle α, scaling it by $k = |A|$, and then translating the object by a vector $T = a + ib$.

2.3 Analytic functions

In this section:

1. The complex derivative.
2. Cauchy–Riemann equations.

3. The real and imaginary parts of analytic complex variable functions are harmonic functions.

Every complex variable function is composed of two real functions; namely, a real part and an imaginary part:

$$\begin{aligned} f(z) &= u(x,y) + iv(x,y) \\ \mathrm{Re}\left[f(z)\right] &= u(x,y) \\ \mathrm{Im}\left[f(z)\right] &= v(x,y). \end{aligned} \tag{2.7}$$

In this section, we will present the most important properties of analytic functions for our later work with the Complex Variable Boundary Element Method in Part II. To be specific, we will cover the following properties:

1. Analytic functions satisfy the Cauchy–Riemann equations.
2. The real and imaginary parts of an analytic complex variable function are conjugate functions.
3. The real and imaginary parts of an analytic complex variable function are harmonic functions.
4. The isocontours of the real and imaginary parts of an analytic complex variable function form orthogonal curve sets. Consequently, when the isocontours of the two parts are superimposed, the result is known as a flow-net, which has many uses in engineering.

The study of analytic functions in the theory of complex variables is of paramount importance. However, before we can deal directly with analytic functions, we must establish the meaning of differentiability for complex functions. To do this we will define the familiar concepts from calculus of limits, continuity, and the derivative within the context of complex variable functions.

In the definitions that follow, we assume that $f(z)$ is defined on open sets of the complex plane. Therefore, if $f(z_0)$ is defined for some $z_0 \in \mathbb{C}$, then there exists $\varepsilon > 0$ such that $f(z)$ is defined for all $|z - z_0| < \varepsilon$.

2.3.1 Limits

Definition 2.3.1.1 Let $f(z)$ denote a complex function defined on an open set $D \subseteq \mathbb{C}$. Let $z_0 \in D$ and $w_0 \in \mathbb{C}$. We say that $\boldsymbol{w_0}$ **is the limit of** $\boldsymbol{f(z)}$ **as** $\boldsymbol{z}$ **approaches** $\boldsymbol{z_0}$ if and only if for each $\varepsilon > 0$, there exists $\delta > 0$ such that $0 < |z - z_0| < \delta$ implies $|f(z) - w_0| < \varepsilon$. When the limit exists, it can be written as:

$$\lim_{z \to z_0} f(z) = w_0$$

Note 2.3.1.1

Recall the definition of modulus from Definition 1.2.2.1:

$$|z| = |a + ib| = \sqrt{a^2 + b^2}$$

Using this definition, we may interpret the statement $0 < |z - z_0| < \delta$ as:

$$0 < |z - z_0| = |(x - x_0) + i(y - y_0)| = \sqrt{(x - x_0)^2 + (y - y_0)^2} < \delta.$$

Likewise, if we let $f(z) = u(x, y) + iv(x, y)$ and $w_0 = u_0 + iv_0$, then we may interpret $|f(z) - w_0| < \varepsilon$ as:

$$\begin{aligned} |f(z) - w_0| &= \left|\left[u(x, y) - u_0\right] + i\left[v(x, y) - v_0\right]\right| \\ &= \sqrt{\left[u(x, y) - u_0\right]^2 + \left[v(x, y) - v_0\right]^2} \\ &< \varepsilon. \end{aligned}$$

Theorem 2.3.1.2 *Let* $f(z) = u(x, y) + iv(x, y)$, $z_0 = x_0 + iy_0$, *and* $w_0 = u_0 + iv_0$. *Then* $\lim_{z \to z_0} f(z) = w_0$ *if and only if:*

$$\lim_{(x,y) \to (x_0, y_0)} u(x, y) = u_0 \quad \textit{and} \quad \lim_{(x,y) \to (x_0, y_0)} v(x, y) = v_0.$$

Theorem 2.3.1.3 *Let* $\lim_{z \to z_0} f(z) = w_0$ *and* $\lim_{z \to z_0} g(z) = w_1$.

Sum: $\lim_{z \to z_0} [f(z) + g(z)] = w_0 + w_1$.

Product: $\lim_{z \to z_0} [f(z)g(z)] = w_0 w_1$.

Quotient: $\lim_{z \to z_0} \frac{f(z)}{g(z)} = \frac{w_0}{w_1}$, *provided* $w_1 \neq 0$.

2.3.2 Continuity

Definition 2.3.2.1 A complex function, $f(z)$, is said to be **continuous at the point z_0** if the conditions in Equations (2.8)–(2.10) are met:

$$\lim_{z \to z_0} f(z) \text{ exists} \tag{2.8}$$

$$f(z_0) \text{ exists} \tag{2.9}$$

$$\lim_{z \to z_0} f(z) = f(z_0) \tag{2.10}$$

We say that **f is continuous on the set D** if f is continuous at all points $z_0 \in D$.

Note 2.3.2.1

The conditions in Equations (2.8) and (2.9) are actually contained in Equation (2.10) because in order for Equation (2.10) to be true, the conditions in Equations (2.8) and (2.9) must both be true.

An equivalent definition of continuity is as follows. A complex function, $f(z)$, is said to be continuous at the point z_0 if for each $\varepsilon > 0$, there exists $\delta > 0$ such that the condition $|z - z_0| < \delta$ implies $|f(z) - f(z_0)| < \varepsilon$.

Theorem 2.3.2.2 *Let $f(z)$ and $g(z)$ denote complex functions. If f and g are both continuous at $z_0 \in \mathbb{C}$, then $f + g$ and $f \cdot g$ are also continuous at z_0. Further, f/g is continuous at z_0 provided that $g(z_0) \neq 0$.*

Note 2.3.2.2

Of the properties included in Theorem 2.3.2.2, the most important for our study is the fact that a sum of continuous functions is itself continuous. As we will discuss in Part II, the approximation function developed using the Complex Variable Boundary Element Method (CVBEM) is a sum of continuous complex functions. Consequently, the CVBEM approximation function is itself a continuous complex function.

Theorem 2.3.2.3 *If $f(z) = u(x, y) + iv(x, y)$ is continuous, then u and v are also continuous.*

Note 2.3.2.3

As explained in Note 2.3.2.2, we are able to conclude that the CVBEM approximation function is a continuous complex function. Therefore, by Theorem 2.3.2.3, we are able to conclude that the real and imaginary parts of a CVBEM approximation function are also continuous. Thus, the CVBEM provides continuous computational estimates of the solution throughout the problem domain.

2.3.3 The derivative

Definition 2.3.3.1 Let $f(z)$ denote a complex function whose domain of definition includes the neighborhood of some point z_0. The **derivative of $f(z)$ at z_0** is defined as the limit:

$$f'(z_0) = \lim_{z \to z_0} \frac{f(z) - f(z_0)}{z - z_0} \tag{2.11}$$

If the limit in Equation (2.11) exists, then f is said to be **differentiable at** z_0.

Note 2.3.3.1

Let $z = x + iy$, $z_0 = x_0 + iy_0$, $f(z) = u(x, y) + iv(x, y)$ and $f(z_0) = u(x_0, y_0) + iv(x_0, y_0)$. Using this notation, Equation (2.11) can be written equivalently as follows:

$$f'(z_0) = \lim_{x+iy \to x_0+iy_0} \frac{[u(x, y) - u(x_0, y_0)] + i[v(x, y) - v(x_0, y_0)]}{[x - x_0] + i[y - y_0]} \tag{2.12}$$

Example 2.3.3.1

Consider the function $f(z) = \alpha z + \beta$, where $\alpha \in \mathbb{R}$ and $\beta \in \mathbb{R}$. We will compute the derivative of this function using Definition 2.3.3.1.

$$f'(z) = \lim_{w \to z} \frac{f(w) - f(z)}{w - z} = \frac{(\alpha w + \beta) - (\alpha z + \beta)}{w - z} = \alpha \frac{w - z}{w - z} = \alpha$$

Many of the rules from real variable calculus for computing the derivative of a single variable function can be applied to computing the derivative of a complex variable function. The following are a few examples:

Function	Complex Derivative
$f(z) = z^n$	$f'(z) = nz^{n-1},\ n \geq 1$
$f(z) = \ln(z)$	$f'(z) = \frac{1}{z},\ z \neq 0$
$f(z) = e^{\alpha z},\ \alpha \in \mathbb{R}$	$f'(z) = \alpha e^{\alpha z}$
$f(z) = \sin z$	$f'(z) = \cos z$
$f(z) = \cos z$	$f'(z) = -\sin z$

Theorem 2.3.3.2 *If f is differentiable at z_0, then f is continuous at z_0.*

Proof. Suppose f is differentiable at z_0. We seek to show that $\lim_{z \to z_0} f(z) = f(z_0)$. Observe,

$$\begin{aligned}
\lim_{z \to z_0} f(z) &= \lim_{z \to z_0} \left[\frac{f(z) - f(z_0)}{z - z_0}(z - z_0) + f(z_0) \right] \\
&= \lim_{z \to z_0} \left[\frac{f(z) - f(z_0)}{z - z_0} \right] \cdot \lim_{z \to z_0} [z - z_0] + f(z_0) \\
&= \lim_{z \to z_0} \left[\frac{f(z) - f(z_0)}{z - z_0} \right] \cdot 0 + f(z_0) \\
&= f(z_0).
\end{aligned} \tag{2.13}$$

Note that in Equation (2.13) we have used the product rule for limits, and we have used the fact that the derivative of f exists at z_0 to conclude that $\lim_{z \to z_0} \left[\frac{f(z) - f(z_0)}{z - z_0} \right] \cdot 0 = 0$. □

Note 2.3.3.2

Let f denote a single variable function of the real variable x. Now, consider the limit of f as $x \to x_0$: $\lim_{x \to x_0} f(x)$. We can evaluate this expression by considering the following two limits:

$$\lim_{x \to x_0^-} f(x) \quad \text{and} \quad \lim_{x \to x_0^+} f(x).$$

The first expression denotes the limit of f as x approaches x_0 from the left. The second expression denotes the limit of f as x approaches x_0 from the right. Importantly, $\lim_{x \to x_0} f(x)$ only exists if $\lim_{x \to x_0^-} f(x) = \lim_{x \to x_0^+} f(x)$. However, when considering the limits of complex functions, such as the limit in Equation (2.11), it is important to note that z can approach z_0 in any direction—not just from the left or right, as in the case of real functions of a single variable.

If $f'(z_0)$ exists, then z can approach z_0 in any direction. Let us consider the case in which $z \to z_0$ along the line $y = y_0$. Thus, Equation (2.12) becomes:

$$\begin{aligned} f'(z_0) &= \lim_{x \to x_0} \frac{\left[u(x, y_0) - u(x_0, y_0)\right] + i\left[v(x, y_0) - v(x_0, y_0)\right]}{x - x_0} \\ &= \lim_{x \to x_0} \frac{\left[u(x, y_0) - u(x_0, y_0)\right]}{x - x_0} + i \lim_{x \to x_0} \frac{\left[v(x, y_0) - v(x_0, y_0)\right]}{x - x_0} \\ &= \frac{\partial u}{\partial x} + i\frac{\partial v}{\partial x}. \end{aligned} \tag{2.14}$$

Now, consider the case in which $z \to z_0$ along the line $x = x_0$. Thus, Equation (2.12) becomes:

$$\begin{aligned} f'(z_0) &= \lim_{y \to y_0} \frac{\left[u(x_0, y) - u(x_0, y_0)\right] + i\left[v(x_0, y) - v(x_0, y_0)\right]}{i(y - y_0)} \\ &= -i \lim_{x \to x_0} \frac{\left[u(x_0, y) - u(x_0, y_0)\right]}{y - y_0} + \lim_{y \to y_0} \frac{\left[v(x_0, y) - v(x_0, y_0)\right]}{y - y_0} \\ &= \frac{\partial v}{\partial y} - i\frac{\partial u}{\partial y} \end{aligned}$$

(2.15)

The results of Equations (2.14) and (2.15) must be equivalent, which implies that their real and imaginary parts must also be equivalent. Hence,

$$\frac{\partial u}{\partial x} + i\frac{\partial v}{\partial x} = \frac{\partial v}{\partial y} - i\frac{\partial u}{\partial y} \Longrightarrow \boxed{\frac{\partial u}{\partial x} = \frac{\partial v}{\partial y}} \text{ and } \boxed{\frac{\partial v}{\partial x} = -\frac{\partial u}{\partial y}}. \tag{2.16}$$

The boxed relationships in Equation (2.16) are known as the Cauchy–Riemann equations. A particularly important property of analytic functions is that they must satisfy this pair of partial differential equations, named after French mathematician A.L. Cauchy (1789–1857) and German mathematician G.F.B. Riemann (1826–1866), who discovered and applied them in the development of complex functions.

The Cauchy–Riemann equations imply that if we know either the real or imaginary parts of an analytic function, then we can determine the other part uniquely up to an additive constant. Consider the example below demonstrating this concept.

Example 2.3.3.2

Suppose $f(z) = u(x,y) + iv(x,y)$ is an analytic complex variable function. If we are given $u(x,y) = y^3 - 3x^2y + 5xy$, then we can determine the corresponding v, up to an additive constant, using the Cauchy–Riemann equations.

We begin by computing the necessary partial derivatives:

$$\frac{\partial u}{\partial y} = 3y^2 - 3x^2 + 5x \quad \text{and} \quad \frac{\partial u}{\partial x} = 5y - 6xy.$$

Then, from the Cauchy–Riemann equations, we obtain:

$$\frac{\partial v}{\partial x} = -3y^2 + 3x^2 - 5x \quad \text{and} \quad \frac{\partial v}{\partial y} = 5y - 6xy. \tag{2.17}$$

Next, we integrate $\frac{\partial v}{\partial x} = -3y^2 + 3x^2 - 5x$ with respect to x to obtain:

$$\begin{aligned} v(x,y) &= \int \frac{\partial v}{\partial x}\,\mathrm{d}x = \int -3y^2 + 3x^2 - 5x\,\mathrm{d}x \\ &= -3xy^2 + x^3 - \frac{5x^2}{2} + g(y). \end{aligned} \tag{2.18}$$

Then, differentiating the result in Equation (2.18) with respect to y, we obtain:

$$\frac{\partial v}{\partial y} = \frac{\partial}{\partial y}\left(-3xy^2 + x^3 + g(y)\right) = -6xy + g'(y).$$

Comparing this result with what was computed in Equation (2.17), it is clear that:

$$5y - 6xy = -6xy + g'(y) \implies g'(y) = 5y.$$

Thus, $g(y) = \frac{5y^2}{2} + c$ for some $c \in \mathbb{R}$. Therefore, returning to the expression for v in Equation (2.18), we conclude:

$$v(x, y) = -3xy^2 + x^3 + \frac{5y^2}{2} + c.$$

The fact that the undetermined constant c appears in our equation for v is the reason why we say that we can only uniquely determine v up to an additive constant from the expression for u when using the Cauchy–Riemann equations.

Example 2.3.3.3

Verify that the function $v(x, y)$ obtained in Example 2.3.3.2 is a harmonic function.

In Example 2.3.3.2, we obtained $v(x, y) = -3xy^2 + x^3 + \frac{5y^2}{2} + c$. To verify that v is a harmonic function, we will see if it satisfies Laplace's equation. We compute the necessary partial derivatives:

$$\frac{\partial^2 v}{\partial x^2} = -5 + 6x$$

$$\frac{\partial^2 v}{\partial y^2} = 5 - 6x.$$

Therefore,

$$\nabla^2 v = \frac{\partial^2 v}{\partial x^2} + \frac{\partial^2 v}{\partial y^2} = (-5 + 6x) + (5 - 6x) = 0.$$

Since v satisfies Laplace's equation, we conclude that v is a harmonic function.

2.3.4 Properties of analytic functions

Definition 2.3.4.1 Let $\Omega \subset \mathbb{C}$ denote an open set. A complex function, $f(z)$, is **analytic in Ω** if it is differentiable at every point in Ω.

When f is analytic in $\mathbb{C}$, the function is said to be **entire**.

Note 2.3.4.1

The term "analytic function" is often used interchangeably with "holomorphic function." It is a significant theorem of complex analysis that all analytic functions are holomorphic, and all holomorphic functions are analytic.

Several important properties of analytic complex variable functions are as follows:

Theorem 2.3.4.2 *a. Let $f(z)$ and $g(z)$ be analytic on $D \subseteq C$. Then, $f(z) + g(z)$ is analytic on D, and the derivative of this sum is computed as: $(f+g)' = f'(z) + g'(z)$.*

b. Let $f(z)$ and $g(z)$ be analytic on $D \subseteq C$. Then, $f(z)g(z)$ is analytic on D, and the derivative of this product is computed as: $(fg)' = f'(z)g(z) + f(z)g'(z)$.

c. Let $f(z)$ and $g(z)$ be analytic on $D \subseteq C$. Then, $\frac{f(z)}{g(z)}$ is analytic on D provided $g(z) \neq 0$, and the derivative of this quotient is computed as: $\left(\frac{f}{g}\right)' = \frac{f'(z)g(z) - f(z)g'(z)}{[g(z)]^2}$.

Note 2.3.4.2

Of the properties included in Theorem 2.3.4.2, the most important for our study is the fact that a sum of analytic functions is itself analytic. As we will discuss in Part II, the approximation function developed using the Complex Variable Boundary Element Method is a sum of analytic complex functions. Consequently, the CVBEM approximation function is itself an analytic complex function.

Note 2.3.4.3

It can be shown that if $f(z) = u(x, y) + iv(x, y)$ is analytic, then u and v will have continuous partial derivatives of all orders. Furthermore, the mixed partial derivatives will be equal; namely, $\frac{\partial^2 u}{\partial x \partial y} = \frac{\partial^2 u}{\partial y \partial x}$ and $\frac{\partial^2 v}{\partial x \partial y} = \frac{\partial^2 v}{\partial y \partial x}$. The details of proving these statements are not necessary for our study, but the following consequence is of the greatest importance: from the Cauchy–Riemann equations, we can show that if f is analytic, then u and v are harmonic functions.

Example 2.3.4.1

We will show that if $f(z) = u(x, y) + iv(x, y)$ is analytic, then u and v are harmonic functions. Recall the Cauchy–Riemann equations:

$$\frac{\partial u}{\partial x} = \frac{\partial v}{\partial y} \tag{2.19}$$

$$\frac{\partial v}{\partial x} = -\frac{\partial u}{\partial y}. \tag{2.20}$$

Compute the partial derivative of Equation (2.19) with respect to x, and compute the partial derivative of Equation (2.20) with respect to y to obtain:

$$\frac{\partial^2 u}{\partial x^2} = \frac{\partial^2 v}{\partial x \partial y} \quad \text{and} \quad \frac{\partial^2 v}{\partial y \partial x} = -\frac{\partial^2 u}{\partial y^2}. \tag{2.21}$$

As discussed in Note 2.3.4.3, the mixed partial derivatives of u and v are equal, hence, $\frac{\partial^2 v}{\partial x \partial y} = \frac{\partial^2 v}{\partial y \partial x}$. Therefore, by combining the expressions in

Equation (2.21), we obtain

$$\nabla^2 u = \frac{\partial^2 u}{\partial x^2} + \frac{\partial^2 u}{\partial y^2} = 0.$$

Thus, we conclude that u satisfies Laplace's equation and is, therefore, a harmonic function.

Using a similar argument, we can show that v also satisfies Laplace's equation. Compute the partial derivative of Equation (2.19) with respect to y, and compute the partial derivative of Equation (2.20) with respect to x to obtain:

$$\frac{\partial^2 u}{\partial y \partial x} = \frac{\partial^2 v}{\partial y^2} \quad \text{and} \quad \frac{\partial^2 v}{\partial x^2} = -\frac{\partial^2 u}{\partial x \partial y}. \tag{2.22}$$

As discussed in Note 2.3.4.3, the mixed partial derivatives of u and v are equal, hence, $\frac{\partial^2 u}{\partial x \partial y} = \frac{\partial^2 u}{\partial y \partial x}$. Therefore, by combining the expressions in Equation (2.22), we obtain

$$\nabla^2 v = \frac{\partial^2 v}{\partial x^2} + \frac{\partial^2 v}{\partial y^2} = 0.$$

Thus, we conclude that v is a harmonic function.

Note 2.3.4.4

As explained in Note 2.3.4.2, we are able to conclude that the CVBEM approximation function is an analytic complex function. Therefore, from the work in Example 2.3.4.1, we are able to conclude that the real and imaginary parts of a CVBEM approximation function are harmonic functions.

The important take-away from this section is that the real and imaginary parts of an analytic complex function are harmonic functions. Therefore, finding real variable functions that satisfy Laplace's equation in two spatial dimensions is as easy as determining the real and imaginary parts of an analytic complex function. This is a central component of modeling using the Complex Variable Boundary Element Method.

2.4 Taylor series representation

In this section:

1. Sequences and series.
2. Convergence properties.
3. Power series.

The Taylor series expansion is important in the study of complex variables because it enables many important computations such as complex integration. First, we will review properties of complex series in general, and then we will discuss Taylor series specifically.

2.4.1 Convergent sequences

Fundamental to the understanding of series is a firm grasp of sequences. Particularly relevant to our discussion is an infinite sequence of complex numbers and the convergence of such a sequence. An infinite sequence is said to converge if a finite limit of the sequence exists, as made clear in the following definition.

Definition 2.4.1.1 An infinite sequence of complex numbers, denoted z_1, $z_2, z_3, \ldots$, **has a limit** $z \in \mathbb{C}$ if for every $\varepsilon > 0$, there exists a positive integer N such that,

$$|z_n - z| < \varepsilon \text{ whenever } n > N. \tag{2.23}$$

If a sequence of complex numbers converges to a limit z, we write

$$\lim_{n\to\infty} z_n = z, \quad \text{or equivalently,} \quad z_n \to z. \tag{2.24}$$

A sequence is said to **diverge** if it has no limit.

Note 2.4.1.1
A sequence can have at most one limit.

Two theorems are particularly relevant for the study of convergent sequences of complex variables. The first makes use of the fact that a complex number can be decomposed into real and imaginary components and allows us to assess whether or not a complex sequence converges.

Theorem 2.4.1.2 *If the real and imaginary parts of a sequence of complex numbers converge, so too does the sequence itself. Let $z_n = x_n + iy_n$, where $n = 1, 2, 3...$, be a sequence of complex numbers, and let $z = x + iy$ be its finite limit. Then,*

$$\lim_{n\to\infty} z_n = z \text{ if and only if } \lim_{n\to\infty} x_n = x \text{ and } \lim_{n\to\infty} y_n = y.$$

The next theorem makes use of Cauchy sequences as another tool to assess the convergence of sequences.

Definition 2.4.1.3 A complex sequence z_n is a **Cauchy sequence** if and only if given any positive number ε there exists a positive index N such that $|z_n - z_m| < \varepsilon$ whenever $n > N$ and $m > N$ where z_m is any term in the sequence after z_n.

Theorem 2.4.1.4 *A sequence of complex numbers is convergent if and only if it is a Cauchy sequence.*

2.4.2 Convergent series

In a similar manner, a series of complex numbers is said to converge if it converges to a sum, $S \in \mathbb{C}$. Consider an infinite complex series:

$$\sum_{n=1}^{\infty} z_n = z_1 + z_2 + \cdots \tag{2.25}$$

This series is said to converge to S if its sequence of partial sums converges to S. The Nth partial sum is as follows:

$$s_N = \sum_{n=1}^{N} z_n = z_1 + z_2 + \cdots + z_N, \quad N \geq 1. \tag{2.26}$$

The set of all partial sums forms a sequence, $(s_N)_{N=1}^{\infty}$, and we can discuss the convergence of Equation (2.25) using the language of sequences. In particular, if $\lim_{N\to\infty} s_N = S$ exists and is finite, then we say the infinite series in Equation (2.25) is convergent and $\sum_{n=1}^{\infty} z_n = S$. Like sequences, the sum of a series is unique and a series is termed divergent if it does not converge. The following theorems are the series-equivalent versions of the two theorems discussed in Section 2.4.1 for sequences.

Theorem 2.4.2.1 *If the real and imaginary parts of a series of complex numbers converge to a sum, so too does the series itself. Let* $z(n) = x_n + iy_n$ *where* $n = 1, 2, 3...$ *be a series of complex numbers and* $s = X + iY$ *be its sum. Then,*

$$\sum_{n=1}^{\infty} z_n = s \text{ if and only if } \sum_{n=1}^{\infty} x_n = X \text{ and } \sum_{n=1}^{\infty} y_n = Y.$$

Definition 2.4.2.2 A complex series, $\sum_{n=1}^{\infty} z_n$, **converges absolutely** if and only if there exists $L \in \mathbb{R}$ such that:

$$\sum_{n=1}^{\infty} |z_n| = \sum_{n=1}^{\infty} \sqrt{x_n^2 + y_n^2} = L \tag{2.27}$$

Theorem 2.4.2.3 *If the complex series* $\sum_{n=1}^{\infty} z_n$ *converges absolutely, then it converges.*

Intuitively, this theorem follows from the fact that absolute convergence is much more difficult to achieve than normal convergence.

2.4.3 Uniform convergence

Definition 2.4.3.1 Consider a sequence of complex functions, $\{f_n\}$. Sequence $\{f_n\}$ **converges to** f **uniformly** on a subset S of domain D if and only if for every positive number ε there is an index N such that $|f(z) - f_n(z)| < \varepsilon$ for all $n > N$ and z in S.

The important aspect of uniform convergence to note is that the index N depends only on the sequence $\{f_n\}$ and ε, not on the point z.

The significance of uniform convergence is its use in proving the continuity of power series functions. This result comes from the following theorem:

Theorem 2.4.3.2 *If* $\{f_n\}$ *is a uniformly convergent sequence of complex numbers, then its limit function,* f*, is itself continuous.*

2.4.4 Power series

A power series is a particular type of series of the form $a_0 + a_1(z - z_0) + a_2(z - z_0)^2 + \ldots$, rewritten more generally as

$$\sum_{n=0}^{\infty} a_n(z - z_0)^n. \tag{2.28}$$

A particularly important power series concept is the radius of convergence, R, which is a non-negative number ($0 \le R < \infty$) that satisfies the following requirements:

1. The series converges absolutely for every z where $|z - z_o| < R$ and at the point $z = z_0$ if $R = 0$. In other words, the series converges for all points in the circle of convergence.

2. The series is unbounded and thus diverges for every z where $|z - z_o| > R$. This is the set of points that lay outside of the circle of convergence.

Now that the power series has been introduced, it is time to identify its relevance to our study of complex variables: convergent power series yield analytic functions. This result arises from Weierstrass' theorem, stated below.

Theorem 2.4.4.1 *Suppose $\{f_n\}$ is a sequence of analytic functions defined on the domain D. Let $\{f_n\}$ converge uniformly to a limit function $f(z)$ on D, or $f(z) = \lim_{n\to\infty} f_n(z)$. Then, f is analytic on D. Furthermore, f_n' converges uniformly to f' on D.*

A significant corollary to this theorem follows:

Corollary 2.4.4.2 *An analytic function is defined by a power series at every point inside the open circle of convergence.*

Together, Weierstrass' theorem and the resulting corollary state that a convergent power series yields an analytic function.

2.4.5 Taylor series

With this unique relationship between power series and analytic functions in mind, we find that every analytic function can be represented by a certain type of power series, namely a Taylor series.

Theorem 2.4.5.1 *Let the function f be analytic in a disk $D = |z - z_0| < R_0$ with center z_0 and radius R_0. Then $f(z)$ can be represented in D as a particular power series, $f(z) = \sum_{n=0}^{\infty} a_n(z - z_0)^n$, with the coefficient $a_n = \frac{f^{(n)}(z_0)}{n!}$, $n = 0, 1, 2, 3, \ldots$ where $f^{(n)}$ denotes the nth derivative of f and $f^{(0)} = f$.*

Rewritten, this expansion of $f(z)$ into a Taylor series about z_0 becomes

$$\begin{aligned} f(z) &= f(z_0) + \frac{f'(z_0)}{1!}(z - z_0) + \frac{f''(z_0)}{2!}(z - z_0)^2 + \cdots \\ &= \sum_{n=0}^{\infty} \frac{f^{(n)}(z_0)}{n!}(z - z_0)^n \end{aligned} \tag{2.29}$$

Any function that is analytic at a point $z_0 \in \mathbb{C}$ has a Taylor series expansion about that point and can be represented using Equation (2.29). However, calculating any more than the first few terms of a Taylor series can be tedious to do by hand. Therefore, in the next section, we will examine how *Excel* can be employed to make this task easier for several commonly encountered functions in complex variables.

2.5 Complex functions

In this section:

1. Common real variable functions and their complex variable analogues.
2. Taylor series approximations of real and complex variable functions.
3. *Excel* implementation of real and complex variable Taylor series.

We now turn our attention to individual complex functions. The Taylor series examined in this section will be useful for becoming more familiar with complex variable functions. Furthermore, practice implementing these series in *Excel* will be helpful later when implementing the CVBEM approximation function, which is itself a series-type approximation function.

The primary effort in this section is to use *Excel* to compute the Taylor series expansions of frequently-encountered complex variable functions. As you read, pay close attention to the *Excel* code used to implement the Taylor series representations of the selected complex functions and the ability of partial sums of these Taylor series to evaluate functions at any point where the series are defined.

Note 2.5.0.1

In this section, we will use the notation T_n to represent the nth partial sum of a Taylor series. Therefore, since the Taylor series begins with the term at $n = 0$, the partial sum T_n requires summing the first $n + 1$ terms in the Taylor series.

2.5.1 The power function

The first function we will consider is a basic power function, $f(z) = z^n$, where n is a non-negative real number. The derivative of a power function is computed using the power rule.

$$\frac{\mathrm{d}}{\mathrm{d}z} z^n = \begin{cases} nz^{n-1}, & n > 0 \\ 0, & n = 0 \end{cases} \qquad n \in \mathbb{R}, \quad n \geq 0 \tag{2.30}$$

Note 2.5.1.1

Power functions are always continuous.

Note 2.5.1.2

Taylor series are a type of power series representation for functions. When negative integer powers are considered, the result is called a Laurent series. Laurent series are not considered power functions.

Note 2.5.1.3

Power series have occasionally been used as basis functions in the CVBEM. When these basis functions are used, the resulting variant of the CVBEM has been called the Complex Polynomial Method [11, 12, 13, 14, 15].

Example 2.5.1.1

Consider the following two situations: raising a complex number to a real exponent, and raising a complex number to a complex exponent.

When raising a complex base, z, to a real power, a, use the following equation where $z = re^{i\theta}$ is the polar form of z:

$$\begin{aligned} z^a &= \left(re^{i\theta}\right)^a \\ &= r^a e^{ia\theta} \\ &= r^a\left[\cos(a\theta) + i\sin(a\theta)\right]. \end{aligned} \tag{2.31}$$

If using both a complex base and a complex power, the equation becomes slightly more complicated. Below is the standard equation for such a situation.

$$(a+ib)^{c+id} = \left(a^2+b^2\right)^{\frac{(c+id)}{2}} e^{i(c+id)\arg(a+ib)}, \quad a,b,c,d \in \mathbb{R} \tag{2.32}$$

***Excel* Implementation 2.5.1.1**

When attempting to compute z^a or z^{a+ib} in *Excel*, the command is IMPOWER. The IMPOWER function takes two arguments: (1) the number to be raised to a power (potentially a complex input), and (2) the power to which the first argument is to be raised. The code that follows would raise a complex number in cell G8 to the 5th power.

=IMPOWER(G8,5) (2.33)

2.5.2 The square root function

The next function we will consider is the square root function, $f(z) = z^{1/2}$. In real variables, the domain of the square root function is $[0, \infty)$. However, in complex variables, the square root function can be evaluated for any complex input. The derivative of the square root function can be computed using the power rule:

$$\frac{\mathrm{d}}{\mathrm{d}z} z^{1/2} = \frac{z^{-1/2}}{2}.$$

In general, the nth derivative of the square root function is calculated by:

$$\frac{\mathrm{d}^n}{\mathrm{d}z^n} z^{1/2} = (-1)^{n+1} \frac{(2n-2)!}{2^{2n-1}(n-1)!} \frac{1}{z^{n-1/2}}, \quad z \neq 0, \quad n \geq 0. \tag{2.34}$$

The Taylor series for $z^{1/2}$ expanded about the point $z_0 = 0$ does not exist because the square root function is not differentiable at $z_0 = 0$. Therefore, we provide the following representation for the translated square root function $\sqrt{1+z}$ expanded about the point $z_0 = 0$, as follows:

$$\sqrt{1+z} = \sum_{n=0}^{\infty} \frac{(-1)^n (2n)!}{(1-2n)(n!)^2 (4^n)} z^n, \quad |z| < 1. \tag{2.35}$$

Tables 2.1–2.2 are provided to demonstrate the *Excel* syntax for computing partial sums of the Taylor series for the real variable translated square root function, $\sqrt{1+x}$. The provided code is for the Taylor series expanded about the point $x_0 = 0$ and evaluated at the point $x = 0.5$. Furthermore, the partial sum approximations $T_0, T_1, \ldots, T_{20}$ for the real variable translated square root function are shown in Figure 2.19.

Excel Implementation 2.5.2.1

Cell	C2
Description	Real Evaluation Point
Code	=0.5
Output	0.5

Table 2.1: This cell contains the real number at which the Taylor series partial sum will be evaluated.

In Figure 2.19, the first column is used to define the value of the index n for the formula given in Equation (2.35). The second and third columns are used to construct each term of Equation (2.35). The fourth column sums all previous terms together to produce the partial sum approximation of $\sqrt{1+0.5}$. The fifth column computes an absolute error of the Taylor series partial sum approximation.

Taylor Series Approximation of Real Square Root Function

ENTER Evaluation Point -----> x =		0.5		
n	**$(-1)^n(2n)!$**	**$(-1)^n(2n)!/((1-2n)(n!)^2(4^n))x^n$**	**Partial Sum Approximation**	**Absolute Error**
0	1.000E+00	1.000E+00	1.000E+00	2.247E-01
1	-2.000E+00	2.500E-01	1.250E+00	2.526E-02
2	2.400E+01	-3.125E-02	1.219E+00	5.995E-03
3	-7.200E+02	7.813E-03	1.227E+00	1.818E-03
4	4.032E+04	-2.441E-03	1.224E+00	6.238E-04
5	-3.629E+06	8.545E-04	1.225E+00	2.307E-04
6	4.790E+08	-3.204E-04	1.225E+00	8.972E-05
7	-8.718E+10	1.259E-04	1.225E+00	3.616E-05
8	2.092E+13	-5.114E-05	1.225E+00	1.498E-05
9	-6.402E+15	2.131E-05	1.225E+00	6.333E-06
10	2.433E+18	-9.056E-06	1.225E+00	2.723E-06
11	-1.124E+21	3.911E-06	1.225E+00	1.187E-06
12	6.204E+23	-1.711E-06	1.225E+00	5.236E-07
13	-4.033E+26	7.567E-07	1.225E+00	2.331E-07
14	3.049E+29	-3.378E-07	1.225E+00	1.047E-07
15	-2.653E+32	1.520E-07	1.225E+00	4.734E-08
16	2.631E+35	-6.889E-08	1.225E+00	2.154E-08
17	-2.952E+38	3.140E-08	1.225E+00	9.859E-09
18	3.720E+41	-1.439E-08	1.225E+00	4.534E-09
19	-5.230E+44	6.629E-09	1.225E+00	2.094E-09
20	8.159E+47	-3.066E-09	1.225E+00	9.713E-10

Figure 2.19: Output of an *Excel* program computing partial sum approximations of the real variable translated square root function at a user-specified evaluation point.

Notice how the absolute errors reported in the fifth column of Figure 2.19 demonstrate that the partial sums of the Taylor series appear to be converging.

Table 2.2 shows the second row of the *Excel* code for the Taylor series approximation of the real variable translated square root function. This row is fully generalized and can be extended downwards to create a partial sum approximation using more terms. The use of more terms will result in a more accurate approximation of the real variable translated square root function. Figure 2.19 displays the output of an *Excel* program computing partial sum approximations of the real variable translated square root function at a user-specified evaluation point (in this case, $x = 0.5$).

Excel Implementation 2.5.2.2

Cell	A5	B5	C5	D5	E5
Description	n	$(-1)^n(2n)!$	$(-1)^n(2n)!/\left[(1-2n)(n!)^2(4^n)\right]x^n$	Partial Sum Approximation	Absolute Error
Code	0	=POWER(−1,A5)* FACT(2*A5)	=B5/((1-2*A5)*POWER(FACT(A5),2)* (POWER(4,A5)))*POWER($C$2,A5)	=SUM($C$5:C5)	=ABS(D5-SQRT(1+$C$2))
Output	0	1.000E+00	1.000E + 00	1.000E + 00	2.247E-01

Cell	A6	B6	C6	D6	E6
Description	n	$(-1)^n(2n)!$	$(-1)^n(2n)!/\left[(1-2n)(n!)^2(4^n)\right]x^n$	Partial Sum Approximation	Absolute Error
Code	=A5+1	=POWER(-1,A6)* FACT(2*A6)	=B6/((1-2*A6)*POWER(FACT(A6),2)* (POWER(4,A6)))*POWER($C$2,A6)	=SUM($C$5:C6)	=ABS(D6-SQRT(1+$C$2))
Output	1	−2.000E + 00	2.500E-01	1.250E + 00	2.526E-02

Table 2.2: This is the second row of the *Excel* code for the Taylor series approximation of the real variable translated square root function. This row is fully generalized and can be extended downwards to create a partial sum approximation using more terms.

Tables 2.3–2.4 show the code and output for the complex variable equivalent of the implementation of the real variable square root function given above. Like the real variable example, extending the code to include more terms in the partial sum will increase the accuracy of the approximation.

Table 2.4 shows the second row of the *Excel* code for the Taylor series approximation of the complex variable square root function. Similar to the real variable example, this is a fully generalized row that can be extended downwards to create partial sum approximations with more terms. Figure 2.20 displays the *Excel* output for partial sum approximations of the complex variable square root function at a user-specified evaluation point (in this case, $0.2 + 0.2i$).

***Excel* Implementation 2.5.2.3**

Cell	C2
Description	Complex Evaluation Point
Code	=COMPLEX(0.2,0.2)
Output	$0.2 + 0.2i$

Table 2.3: This cell contains the complex number at which the Taylor series partial sum will be evaluated.

Excel Implementation 2.5.2.4

Cell	A5	B5	C5	D5	E5
Description	n	$(-1)^n(2n)!$	$(-1)^n(2n)!/\left[(1-2n)(n!)^2(4^n)\right]z^n$	Partial Sum Approximation	Absolute Error
Code	0	=POWER(-1,A5)* FACT(2*A5)	=IMPRODUCT(B5/IMPRODUCT((1-2*A5),IMPOWER(FACT(A5),2), (IMPOWER(4,A5))),IMPOWER($C$2,A5))	=SUM($C$5:C5)	=IMABS(IMSUB(D5, IMSQRT(IMSUM(1,$C$2))))
Output	0	1.000E+00	1	1	1.356E-01

Cell	A6	B6	C6	D6	E6
Description	n	$(-1)^n(2n)!$	$(-1)^n(2n)!/\left[(1-2n)(n!)^2(4^n)\right]z^n$	Partial Sum Approximation	Absolute Error
Code	=A5+1	=POWER(-1,A6)* FACT(2*A6)	=IMPRODUCT(B6/IMPRODUCT((1-2*A6),IMPOWER(FACT(A6),2), (IMPOWER(4,A6))),IMPOWER($C$2,A6))	=SUM($C$5:C6)	=IMABS(IMSUB(D6, IMSQRT(IMSUM(1,*C*110))))
Output	1	-2.000E+00	$0.1+0.1i$	$1.1+0.1i$	9.060E-03

Table 2.4: This is the second row of the *Excel* code for the Taylor series approximation of the complex variable translated square root function. This row is fully generalized and can be extended downwards to create a partial sum approximation using more terms.

Taylor Series Approximation of Complex Square Root Function

ENTER Evaluation Point ------> z=		0.2+0.2i		
n	**(-1)^n(2n)!**	**(-1)^n(2n)!/((1-2n)(n!)^2(4^n))z^n**	**Partial Sum Approximation**	**Absolute Error**
0	1.000E+00	1	1	1.346E-01
1	-2.000E+00	0.1+0.1i	1.1+0.1i	9.060E-03
2	2.400E+01	-6.1257422745431E-19-0.01i	1.1+0.09i	1.250E-03
3	-7.200E+02	-0.001+0.001i	1.099+0.091i	2.178E-04
4	4.032E+04	0.00025-3.06287113727155E-20i	1.09925+0.091i	4.270E-05
5	-3.629E+06	-0.000035-0.000035i	1.099215+0.090965i	8.994E-06
6	4.790E+08	1.92960881648108E-21+0.0000105i	1.099215+0.0909755i	1.988E-06
7	-8.718E+10	0.00000165-0.00000165i	1.09921665+0.09097385i	4.551E-07
8	2.092E+13	-5.36250000000001E-07+1.31397171788949E-22i	1.09921611375+0.09097385i	1.069E-07
9	-6.402E+15	8.93750000000003E-08+8.93750000000002E-08i	1.099216203125+0.090973939375i	2.564E-08
10	2.433E+18	-9.30729966838392E-24-3.03875000000001E-08i	1.099216203125+0.0909739089875i	6.249E-09
11	-1.124E+21	-5.24875000000002E-09+5.24875000000003E-09i	1.0992161978625+0.09097391423625i	1.544E-09
12	6.204E+23	1.83706250000001E-09-6.75202285033671E-25i	1.09921619971331+0.09097391423625i	3.856E-10
13	-4.033E+26	-3.25018750000001E-10-3.25018750000001E-10i	1.09921619938829+0.0909739139112312i	9.726E-11
14	3.049E+29	4.9774527422354E-26+1.16078125000001E-10i	1.09921619938829+0.0909739140273093i	2.473E-11
15	-2.653E+32	2.08940625000001E-11-2.08940625000001E-11i	1.09921619940918+0.0909739140064152i	6.324E-12
16	2.631E+35	-7.57409765625004E-12+3.71175761635269E-27i	1.09921619940161+0.0909739140064152i	1.634E-12
17	-2.952E+38	1.38115898437501E-12+1.38115898437501E-12i	1.09921619940299+0.0909739140077964i	4.218E-13
18	3.720E+41	-2.79200591288882E-28-5.064249609375E-13i	1.09921619940299+0.09097391400729i	1.074E-13
19	-5.230E+44	-9.32888085937507E-14+9.32888085937507E-14i	1.0992161994029+0.0909739140073833i	3.030E-14
20	8.159E+47	3.45168591796874E-14-2.11441383461463E-29i	1.09921619940293+0.0909739140073833i	4.399E-15

Figure 2.20: Output of an *Excel* program computing partial sum approximations of the complex variable translated square root function at a user-specified evaluation point.

Figure 2.21: Contour plot of the real part of sqrt$(z+1)$ for $x \in [0,1]$, $y \in [0,1]$. The contour plot was produced in *Excel* using the instructions given in Section 5.1.5.

Figure 2.22: Contour plot of the imaginary part of sqrt$(z+1)$ for $x \in [0,1]$, $y \in [0,1]$. The contour plot was produced in *Excel* using the instructions given in Section 5.1.5.

2.5.3 The exponential function

The exponential function is a commonly used complex function, and it has some unique properties.

Definition 2.5.3.1 The **complex exponential function** is defined by

$$f(z) = e^z = e^{x+iy} = e^x e^{iy} = e^x (\cos y + i \sin y), \qquad (2.36)$$

where the identity $e^{iy} = (\cos y + i \sin y)$ is known as Euler's identity.

The real and imaginary parts of the complex exponential function are as follows:

$$\begin{aligned} \mathrm{Re}(e^z) &= e^x \cos y \\ \mathrm{Im}(e^z) &= e^x \sin y. \end{aligned} \qquad (2.37)$$

Several important properties of the complex exponential function are as follows. All of the properties have real variable equivalents.

Property 1: $e^{z_1} e^{z_2} = e^{z_1+z_2}$
Property 2: $\frac{e^{z_1}}{e^{z_2}} = e^{z_1-z_2}$

Property 3: $\frac{\mathrm{d}^n}{\mathrm{d}z^n} e^z = e^z$ for all $z \in \mathbb{C}$ and all non-negative integers, n.
Property 4: e^z is a continuous function.
Property 5: $|e^z| = |e^x|$:

$$\begin{aligned} |e^z| &= \left|e^{x+iy}\right| = \left|e^x e^{iy}\right| \\ &= \left|e^x \left[\cos(y) + i\sin(y)\right]\right| \\ &= \left|e^x \cos(y) + ie^x \sin(y)\right| \\ &= \sqrt{\left[e^x \cos(y)\right]^2 + \left[e^x \sin(y)\right]^2} \\ &= \sqrt{e^{2x} \cos^2(y) + e^{2x} \sin^2(y)} \\ &= \sqrt{e^{2x}\left[\cos^2(y) + \sin^2(y)\right]} \\ &= \sqrt{e^{2x}} \\ &= |e^x| \end{aligned} \tag{2.38}$$

Property 6: $e^0 = 1$.
Property 7: $e^z \neq 0$ for all $z \in \mathbb{C}$.

Note 2.5.3.1

The intuition for Property 7 is as follows. If a complex number is 0, then both the real and imaginary parts of the number must be zero. Therefore, if $e^z = 0$, then we must have simultaneously:

$$e^x \cos y = 0 \quad \text{and} \quad e^x \sin y = 0. \tag{2.39}$$

Since e^x is never 0, this means we must have $\cos y = \sin y = 0$. However, there is no real value of y satisfying this equation. Therefore, we conclude $e^z \neq 0$ for any value of $z \in \mathbb{C}$.

The Taylor series expansion of $f(z) = e^z$ about $z_0 = 0$ where $-\infty < z < \infty$ is

$$e^z = \sum_{n=0}^{\infty} \frac{z^n}{n!} = 1 + \frac{z}{1!} + \frac{z^2}{2!} + \cdots \tag{2.40}$$

Example 2.5.3.1

We will derive the Taylor series for e^z expanded about the point $z_0 = 0$. Let $f(z) = e^z$, and recall the definition of the Taylor series expanded about the point z_0:

$$f(z) = \sum_{n=0}^{\infty} \frac{f^{(n)}(z_0)}{n!}(z - z_0)^n. \tag{2.41}$$

Since we are expanding about $z_0 = 0$, this equation becomes:

$$f(z) = \sum_{n=0}^{\infty} \frac{f^{(n)}(0)}{n!} z^n. \tag{2.42}$$

Using Property 3 above, we have: $\frac{\mathrm{d}^n}{\mathrm{d}z^n} e^z = e^z$. Thus,

$$e^z = \sum_{n=0}^{\infty} \frac{e^z\big|_{z=0}}{n!} z^n. \tag{2.43}$$

Using Property 6 above, we have: $e^0 = 1$. Thus,

$$e^z = \sum_{n=0}^{\infty} \frac{1}{n!} z^n = \sum_{n=0}^{\infty} \frac{z^n}{n!}. \tag{2.44}$$

Tables 2.5–2.7 are provided to demonstrate the *Excel* syntax for computing partial sums of the Taylor series for the real variable exponential function. The provided code is for the Taylor series expanded about the point $x_0 = 0$ evaluated at the point $x = 5$. Furthermore, the partial sum approximations $T_0, T_1, \ldots, T_{20}$ for the real variable exponential function are shown in Figure 2.23.

In Figure 2.23, the first column is used to define the value of the index n for the formula given in Equation (2.40). The second and third columns are used to construct each term of Equation (2.40). The fourth column sums all previous terms together to produce the partial sum approximation of e^5. The fifth column computes an absolute error of the Taylor series partial sum approximation.

Notice how the absolute errors reported in the fifth column of Figure 2.23 demonstrate that the partial sums of the Taylor series appear to be converging.

Table 2.7 shows the second row of the *Excel* code for the Taylor series approximation of the real variable exponential function. This row is fully generalized and can be extended downwards to create a partial sum approximation using more terms. The use of more terms will result in a more accurate approximation of the real variable exponential function. Figure 2.23 displays the output of an *Excel* program computing partial sum approximations of the real variable exponential function at a user-specified evaluation point (in this case, $x = 5$).

Excel Implementation 2.5.3.1

Cell	C2
Description	Real Evaluation Point
Code	5
Output	5

Table 2.5: This cell contains the real number at which the Taylor series partial sum will be evaluated.

Excel Implementation 2.5.3.2

Cell	A5	B5	C5
Description	n	$n!$	$x^n/n!$
Code	0	=FACT(A5)	=POWER($C$2,A5)/B5
Output	0	1.000E+00	1.000E+00

Cell	D5	E5
Description	Partial Sum Approximation	Absolute Error
Code	=SUM($C$5:C5)	=ABS(D5-EXP($C$2))
Output	1.000E + 00	1.474E + 02

Table 2.6: This is the first row of the *Excel* code for the Taylor series approximation of the real variable exponential function.

Tables 2.8–2.10 show the code and output for the complex variable equivalent of the implementation of the real variable exponential function given above. Like the real variable example, extending the code to include more terms in the partial sum will increase the accuracy of the approximation.

Excel Implementation 2.5.3.3

Cell	A6	B6	C6
Description	n	$n!$	$x^n/n!$
Code	=A5+1	=FACT(A6)	=POWER($C$2,A6)/B6
Output	1	1.000E+00	5.000E+00

Cell	D6	E6
Description	Partial Sum Approximation	Absolute Error
Code	=SUM($C$5:C6)	=ABS(D6-EXP($C$2))
Output	6.000E+00	1.424E+02

Table 2.7: This is the second row of the *Excel* code for the Taylor series approximation of the real variable exponential function. This row is fully generalized and can be extended downwards to create a partial sum approximation using more terms.

Taylor Series Approximation of Real Exponential Function

ENTER Evaluation Point -----> x = 5

n	n!	$x^n/n!$	Partial Sum Approximation	Absolute Error
0	1.000E+00	1.000E+00	1.000E+00	1.474E+02
1	1.000E+00	5.000E+00	6.000E+00	1.424E+02
2	2.000E+00	1.250E+01	1.850E+01	1.299E+02
3	6.000E+00	2.083E+01	3.933E+01	1.091E+02
4	2.400E+01	2.604E+01	6.538E+01	8.304E+01
5	1.200E+02	2.604E+01	9.142E+01	5.700E+01
6	7.200E+02	2.170E+01	1.131E+02	3.530E+01
7	5.040E+03	1.550E+01	1.286E+02	1.979E+01
8	4.032E+04	9.688E+00	1.383E+02	1.011E+01
9	3.629E+05	5.382E+00	1.437E+02	4.724E+00
10	3.629E+06	2.691E+00	1.464E+02	2.033E+00
11	3.992E+07	1.223E+00	1.476E+02	8.093E-01
12	4.790E+08	5.097E-01	1.481E+02	2.996E-01
13	6.227E+09	1.960E-01	1.483E+02	1.036E-01
14	8.718E+10	7.001E-02	1.484E+02	3.358E-02
15	1.308E+12	2.334E-02	1.484E+02	1.024E-02
16	2.092E+13	7.293E-03	1.484E+02	2.949E-03
17	3.557E+14	2.145E-03	1.484E+02	8.039E-04
18	6.402E+15	5.958E-04	1.484E+02	2.080E-04
19	1.216E+17	1.568E-04	1.484E+02	5.123E-05
20	2.433E+18	3.920E-05	1.484E+02	1.204E-05

Figure 2.23: Output of an *Excel* program computing partial sum approximations of the real variable exponential function at a user-specified evaluation point.

Excel Implementation 2.5.3.4

Cell	C2
Description	Complex Evaluation Point
Code	=COMPLEX(2,2)
Output	$2 + 2i$

Table 2.8: This cell contains the complex number at which the Taylor series partial sum will be evaluated.

***Excel* Implementation 2.5.3.5**

Cell	A5	B5	C5
Description	n	$n!$	$z^n/n!$
Code	0	=FACT(A5)	=IMDIV(IMPOWER ($C$2,A5),B5)
Output	0	1.000E+00	1

Cell	D5	E5
Description	Partial Sum Approximation	Absolute Error
Code	=IMSUM($C$5:C5)	=IMABS(IMSUB(D5, IMEXP($C$2)))
Output	1	7.858E+00

Table 2.9: This is the first row of the *Excel* code for the Taylor series approximation of the complex variable exponential function.

Table 2.10 shows the second row of the *Excel* code for the Taylor series approximation of the complex variable exponential function. Similar to the real variable example, this is a fully generalized row that can be extended downwards to create partial sum approximations with more terms. Figure 2.24 displays the *Excel* output for partial sum approximations of the complex variable exponential function at a user-specified evaluation point (in this case, $2 + 2i$).

Excel Implementation 2.5.3.6

Cell	A6	B6	C6
Description	n	$n!$	$z^n/n!$
Code	=A5+1	=FACT(A6)	=IMDIV(IMPOWER ($C$2,A6),B6)
Output	1	1.000E+00	$2 + 2i$

Cell	D6	E6
Description	Partial Sum Approximation	Absolute Error
Code	=IMSUM($C$5:C6)	=IMABS(IMSUB (D6,IMEXP($C$2)))
Output	$3 + 2i$	7.692E+00

Table 2.10: This is the second row of the *Excel* code for the Taylor series approximation of the complex variable exponential function. This row is fully generalized and can be extended downwards to create a partial sum approximation using more terms.

Taylor Series Approximation of Complex Exponential Function

ENTER Evaluation Point ----> z = 2+2i

n	n!	$z^n/n!$	Partial Sum Approximation	Absolute Error
0	1.000E+00	1	1	7.858E+00
1	1.000E+00	2+2i	3+2i	7.692E+00
2	2.000E+00	2.45029690981724E-16+4i	3+6i	6.117E+00
3	6.000E+00	-2.66666666666667+2.66666666666667i	0.333333333333333+8.66666666666667i	3.926E+00
4	2.400E+01	-2.66666666666667+3.26706254642299E-16i	-2.33333333333334+8.66666666666667i	2.084E+00
5	1.200E+02	-1.06666666666667-1.06666666666667i	-3.40000000000001+7.6i	9.392E-01
6	7.200E+02	-1.3068250185692E-16-0.711111111111111i	-3.40000000000001+6.88888888888889i	3.669E-01
7	5.040E+03	0.203174603174603-0.203174603174603i	-3.19682539682541+6.68571428571429i	1.263E-01
8	4.032E+04	0.101587301587302-2.48919051156037E-17i	-3.0952380952381+6.68571428571429i	3.886E-02
9	3.629E+05	0.0225749559082893+0.0225749559082893i	-3.07266313932982+6.70828924162257i	1.080E-02
10	3.629E+06	2.76576723506707E-18+0.0090299823633157i	-3.07266313932982+6.71731922398589i	2.737E-03
11	3.992E+07	-0.00164181497514831+0.00164181497514831i	-3.07430495430496+6.71896103896104i	6.372E-04
12	4.790E+08	-0.00054727165838277+2.01146708004879E-19i	-3.07485222596335+6.71896103896104i	1.372E-04
13	6.227E+09	-0.000084195639751195З-0.0000841956397511955i	-3.0749364216031+6.71887684332129i	2.745E-05
14	8.718E+10	-1.0315215795122E-20-0.0000240558970717701i	-3.0749364216031+6.71885278742422i	5.135E-06
15	1.308E+12	3.20745294290268E-06-3.20745294290269E-06i	-3.07493321415016+6.71884957997127i	9.012E-07
16	2.092E+13	8.0186323572567E-07-3.92960601718933E-22i	-3.07493241228692+6.71884957997127i	1.490E-07
17	3.557E+14	9.43368512618435E-08+9.43368512618438E-08i	-3.07493231795007+6.71884967430812i	2.328E-08
18	6.402E+15	1.15576647564392E-23+2.09637447248541E-08i	-3.07493231795007+6.71884969527187i	3.447E-09
19	1.216E+17	-2.20670997103727E-09+2.20670997103728E-09i	-3.07493232015678+6.71884969747858i	4.852E-10
20	2.433E+18	-4.41341994207455E-10+2.70354731144777E-25i	-3.07493232059812+6.71884969747858i	6.507E-11

Figure 2.24: Output of an *Excel* program computing partial sum approximations of the complex variable exponential function at a user-specified evaluation point.

Figure 2.25: Contour plot of the real part of $\exp(z)$ for $x \in [0,1]$, $y \in [0,1]$. The contour plot was produced in *Excel* using the instructions given in Section 5.1.5.

Figure 2.26: Contour plot of the imaginary part of $\exp(z)$ for $x \in [0, 1]$, $y \in [0, 1]$. The contour plot was produced in *Excel* using the instructions given in Section 5.1.5.

2.5.4 The logarithmic function

Definition 2.5.4.1 The **complex logarithmic function**, denoted $\log(z)$, is used to determine the value $w = u + iv$ satisfying $e^w = z$ for a nonzero $z = x + iy = re^{i\theta}$, where $r = |z|$ and $\theta = \arg(z)$ such that $\theta \in (-\pi, \pi]$.

$$\log(z) = \ln(r) + i(\theta + 2n\pi) = \ln|z| + i(\theta + 2n\pi), \quad n \in \mathbb{Z} \tag{2.45}$$

Equation (2.45) is derived as follows:

$$e^{u+iv} = x + iy \implies e^u e^{iv} = x + iy = re^{i\theta}. \tag{2.46}$$

Recall, $e^{i\theta} = e^{i(\theta+2n\pi)}$ for any integer n. Thus, we have: $e^u e^{iv} = re^{i(\theta+2n\pi)}$, from which we conclude $r = e^u$ and $v = \theta + 2n\pi$. Since $z \neq 0$, it follows that $r \neq 0$. Hence, the quantity $u = \ln(r)$ is defined. Therefore,

$$w = u + iv = \ln(r) + i(\theta + 2n\pi), \quad n \in \mathbb{Z}. \tag{2.47}$$

Note 2.5.4.1

In particular, for nonzero z, the logarithmic function satisfies:

$$e^{\log(z)} = z, \quad z \neq 0. \tag{2.48}$$

Note 2.5.4.2

The logarithm function is considered multiple-valued because $\log(z) = \ln(r) + i(\theta + 2n\pi)$ is valid for each $n \in \mathbb{Z}$.

In order to address the fact that $\log(z)$ is multiple-valued, we define the principal value, as follows.

Definition 2.5.4.2 Recall the definition of $\log(z) = \ln(r) + i(\theta + 2n\pi)$. The **principal value of $\log(z)$**, denoted $\mathrm{Log}(z)$, is obtained by setting $n = 0$ in Equation (2.45). Thus,

$$\mathrm{Log}(z) = \ln(r) + i\theta, \quad \theta \in (-\pi, \pi] \tag{2.49}$$

Note 2.5.4.3

When the notation $\log(z)$ appears without specifying any value of n, it is usually safe to assume the principal value, $\mathrm{Log}(z)$, is intended.

The Taylor series expansion of the logarithmic function about $z_0 = 1$ is presented below:

$$\begin{aligned} \ln z &= \sum_{n=1}^{\infty} \frac{(-1)^{n+1}}{n}(z-1)^n \\ &= \frac{(z-1)}{1} - \frac{(z-1)^2}{2} + \frac{(z-1)^3}{3} - \ldots, \quad |z-1| < 1. \end{aligned} \tag{2.50}$$

In general, the nth derivative of the natural logarithm function is calculated by:

$$\frac{\mathrm{d}^n}{\mathrm{d}z^n}\ln(z) = \frac{(-1)^{n+1}(n-1)!}{z^n}. \tag{2.51}$$

Lastly, we discuss branches of the logarithmic function.

Definition 2.5.4.3 Recall that $\log(z) = \ln(r) + i(\theta + 2n\pi)$ is valid for any $n \in \mathbb{Z}$, where $\theta \in (-\pi, \pi]$. A **branch** of the logarithm function can be defined by restricting the values of θ to an interval of width 2π. For each $\alpha \in \mathbb{R}$, a branch of the logarithm function is as follows.

$$\log_\alpha(z) = \ln(r) + i\theta_\alpha, \quad \theta_\alpha \in (\alpha, \alpha + 2\pi), \quad r = |z| \tag{2.52}$$

Importantly, the logarithm function is single-valued and analytic throughout the domain $r > 0$ and $\theta \in (\alpha, \alpha + 2\pi)$.

The importance of defining a branch of the logarithm function is that it makes the function single-valued.

Note 2.5.4.4

In Equation (2.52), the logarithm function is not defined on the ray $\theta = \alpha$. The logarithm function would be discontinuous if defined along the ray $\theta = \alpha$.

Definition 2.5.4.4 A **branch cut** of a multiple-valued function is a line or curve that is introduced to define a particular branch of the function. The points on the branch cut are singular points.

Note 2.5.4.5

In the case of the complex logarithm, the branch cut is given by the ray $\theta = \alpha$.

Example 2.5.4.1

Recall the definition of the complex logarithm given above: $\log(z) = \ln(r) + i(\theta + 2n\pi) = \ln|z| + i(\theta + 2n\pi)$, $n \in \mathbb{Z}$, where $\theta \in (-\pi, \pi]$. Rotating the branch cut is accomplished by redefining the interval of the angle to be $\theta_\alpha \in (\alpha, \alpha + 2\pi]$. Suppose $\alpha = \pi/2$. Thus, we are interested in mapping the interval $(-\pi, \pi]$ to $(\pi/2, \pi/2 + 2\pi]$. This is accomplished by the following:

$$\theta_{\pi/2} = \mod\left(\theta - \frac{\pi}{2}, 2\pi\right) + \frac{\pi}{2}.$$

This mapping is visualized below:

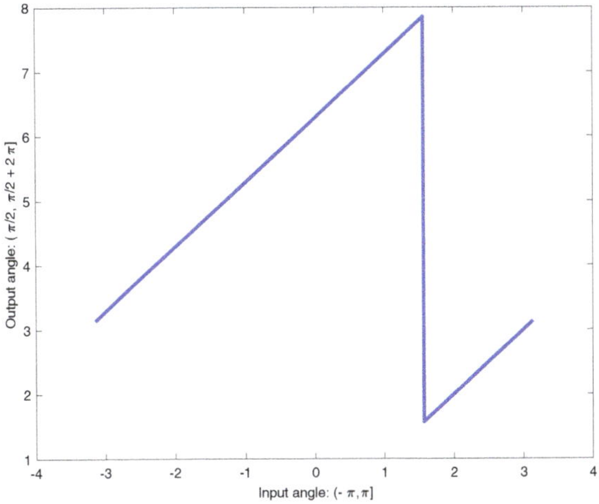

Then, the complex logarithm function with rotated branch cut is as follows:

$$\omega_\alpha(z) = \ln|z| + i\theta_\alpha.$$

And in the specific case of $\alpha = \pi/2$, we have:

$$\begin{aligned}\omega_{\pi/2}(z) &= \ln|z| + i\theta_{\pi/2} \\ &= \ln|z| + i\left[\mod\left(\theta - \frac{\pi}{2}, 2\pi\right) + \frac{\pi}{2}\right].\end{aligned}$$

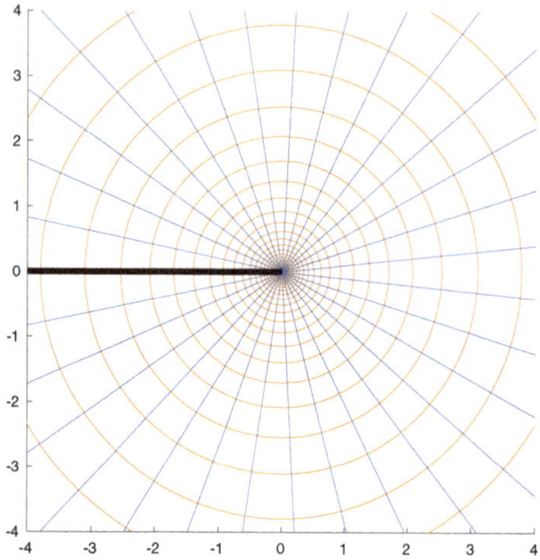

Figure 2.27: Flow net of complex logarithm function with no rotation of the branch cut.

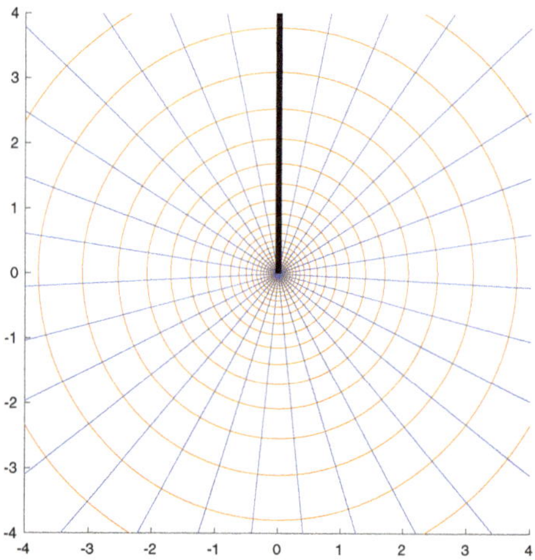

Figure 2.28: Flow net of complex logarithm function with branch cut at angle $\pi/2$.

Tables 2.11–2.12 are provided to demonstrate the *Excel* syntax for computing partial sums of the Taylor series for the real variable logarithmic function. The provided code is for the Taylor series expanded about the point $x_0 = 1$ evaluated at the point $x = 1.5$. Furthermore, the partial sum approximations $T_0, T_1, \ldots, T_{20}$ for the real variable logarithm function are shown in Figure 2.29.

In Figure 2.29, the first column is used to define the value of the index n for the formula given in Equation (2.50). The second and third columns are used to construct each term of Equation (2.50). The fourth column sums all previous terms together to produce the partial sum approximation of $\ln(1.5)$. The fifth column computes an absolute error of the Taylor series partial sum approximation.

Notice how the absolute errors reported in the fifth column of Figure 2.29 demonstrate that the partial sums of the Taylor series appear to be converging.

Table 2.12 shows the second row of the *Excel* code for the Taylor series approximation of the real variable logarithmic function. This row is fully generalized and can be extended downwards to create a partial sum approximation using more terms. The use of more terms will result in a more accurate approximation of the real variable logarithmic function. Figure 2.29 displays the output of an *Excel* program computing partial sum approximations of the real variable logarithmic function at a user-specified evaluation point (in this case, $x = 1.5$).

Excel Implementation 2.5.4.1

Cell	C2
Description	Real Evaluation Point
Code	=1.5
Output	1.5

Table 2.11: This cell contains the real number at which the Taylor series partial sum will be evaluated.

Excel Implementation 2.5.4.2

Cell	A5	B5	C5	D5	E5
Description	n	$(-1)^{n+1}$	$(-1)^{n+1}(x-1)^n/n$	Partial Sum Approximation	Absolute Error
Code	0	=POWER(−1,A5+1)	=B5*POWER($C$2 - 1,A5)/A5	=SUM($C$5:C5)	=IMABS(D5-LN($C$2))
Output	0	1.000E+00	0.5	0.5	9.453E-02

Cell	A6	B6	C6	D6	E6
Description	n	$(-1)^{n+1}$	$(-1)^{n+1}(x-1)^n/n$	Partial Sum Approximation	Absolute Error
Code	=A5+1	=POWER(-1,A6+1)	=B6*POWER($C$2 - 1,A6)/A6	=SUM($C$5:C6)	=IMABS(D6-LN($C$2))
Output	1	−1.000E + 00	−0.125	0.375	3.047E-02

Table 2.12: This is the second row of the *Excel* code for the Taylor series approximation of the real variable logarithmic function. This row is fully generalized and can be extended downwards to create a partial sum approximation using more terms.

Taylor Series Approximation of Real Logarithm Function

ENTER Evaluation Point -----> x = 1.5

n	$(-1)^{n+1}$	$(-1)^{n+1}(x-1)^n/n$	Partial Sum Approximation	Absolute Error
1	1.000E+00	0.5	0.5	9.453E-02
2	-1.000E+00	-0.125	0.375	3.047E-02
3	1.000E+00	0.041666667	0.416666667	1.120E-02
4	-1.000E+00	-0.015625	0.401041667	4.423E-03
5	1.000E+00	0.00625	0.407291667	1.827E-03
6	-1.000E+00	-0.002604167	0.4046875	7.776E-04
7	1.000E+00	0.001116071	0.405803571	3.385E-04
8	-1.000E+00	-0.000488281	0.40531529	1.498E-04
9	1.000E+00	0.000217014	0.405532304	6.720E-05
10	-1.000E+00	-9.76563E-05	0.405434648	3.046E-05
11	1.000E+00	4.43892E-05	0.405479037	1.393E-05
12	-1.000E+00	-2.03451E-05	0.405458692	6.416E-06
13	1.000E+00	9.39002E-06	0.405468082	2.974E-06
14	-1.000E+00	-4.35965E-06	0.405463722	1.386E-06
15	1.000E+00	2.03451E-06	0.405465757	6.487E-07
16	-1.000E+00	-9.53674E-07	0.405464803	3.049E-07
17	1.000E+00	4.48788E-07	0.405465252	1.439E-07
18	-1.000E+00	-2.11928E-07	0.40546504	6.808E-08
19	1.000E+00	1.00387E-07	0.40546514	3.231E-08
20	-1.000E+00	-4.76837E-08	0.405465093	1.537E-08
21	1.000E+00	2.27065E-08	0.405465115	7.333E-09

Figure 2.29: Output of an *Excel* program computing partial sum approximations of the real variable logarithmic function at a user-specified evaluation point.

Tables 2.13–2.14 show the code and output for the complex variable equivalent of the implementation of the real variable logarithmic function given above. Like the real variable example, extending the code to include more terms in the partial sum will increase the accuracy of the approximation.

Table 2.14 shows the second row of the *Excel* code for the Taylor series approximation of the complex variable logarithmic function. Similar to the real variable example, this is a fully generalized row that can be extended downwards to create partial sum approximations with more terms. Figure 2.30 displays the *Excel* output for partial sum approximations of the complex variable logarithmic function at a user-specified evaluation point (in this case, $1.5 + 0.1i$).

Excel Implementation 2.5.4.3

Cell	C2
Description	Complex Evaluation Point
Code	=COMPLEX(1.5,0.1)
Output	$1.5 + 0.1i$

Table 2.13: This cell contains the complex number at which the Taylor series partial sum will be evaluated.

Excel Implementation 2.5.4.4

Cell	A5	B5	C5	D5	E5
Description	n	$(-1)^{n+1}$	$(-1)^{n+1}(z-1)^n/n$	Partial Sum Approximation	Absolute Error
Code	0	=POWER(−1,A5 + 1)	=IMPRODUCT(B5,IMDIV(IMPOWER(IMSUB($C$2,1) ,A5),A5))	=IMSUM($C$5:C5)	=IMABS(IMSUB(D5, IMLN($C$2)))
Output	0	1.000E + 00	$0.5 + 0.1i$	$0.5 + 0.1i$	9.818E-02

Cell	A6	B6	C6	D6	E6
Description	n	$(-1)^{n+1}$	$(-1)^{n+1}(z-1)^n/n$	Partial Sum Approximation	Absolute Error
Code	=A5+1	=POWER(-1,A6+1)	=IMPRODUCT(B6,IMDIV(IMPOWER(IMSUB($C$2,1) ,A6),A6))	=IMSUM($C$5:C6)	=IMABS(IMSUB(D6, IMLN($C$2)))
Output	1	−1.000E + 00	$-0.12 + 0.05i$	$0.38 + 0.05i$	3.226E-02

Table 2.14: This is the second row of the *Excel* code for the Taylor series approximation of the complex variable logarithmic function. This row is fully generalized and can be extended downwards to create a partial sum approximation using more terms.

Taylor Series Approximation of Complex Logarithm Function

ENTER Evaluation Point -----> z = 1.5+0.1i

n	$(-1)^{n+1}$	$(-1)^{n+1}(z-1)^n/n$	Partial Sum Approximation	Absolute Error
1	1.000E+00	0.5+0.1i	0.5+0.1i	9.818E-02
2	-1.000E+00	-0.12-0.05i	0.38+0.05i	3.226E-02
3	1.000E+00	0.0366666666666667+0.0246666666666667i	0.416666666666667+0.0746666666666667	1.210E-02
4	-1.000E+00	-0.0119-0.012i	0.404766666666667+0.0626666666666667	4.871E-03
5	1.000E+00	0.0038+0.005752i	0.408566666666667+0.0684186666666667	2.051E-03
6	-1.000E+00	-0.001104-0.00271333333333333i	0.407462666666667+0.0657053333333334	8.904E-04
7	1.000E+00	0.000240571428571429+0.00125748571428571i	0.407703238095238+0.0669628190476191	3.952E-04
8	-1.000E+00	4.78000000000008E-06-0.0005712i	0.407708018095238+0.0663916190476191	1.784E-04
9	1.000E+00	-0.0000528977777777779+0.000253441777777778i	0.40765512031746+0.0666450608253969i	8.159E-05
10	-1.000E+00	0.00004661376-0.000109288i	0.40770173407746+0.0665357728253969i	3.772E-05
11	1.000E+00	-0.0000311233454545455+0.0000454387490909091i	0.407670610732006+0.0665812115744878	1.759E-05
12	-1.000E+00	0.0000184300853333333-0.00001797312i	0.407689040817339+0.0665632384544878	8.263E-06
13	1.000E+00	-0.0000101652504615385+6.59404750769231E-06i	0.407678875566878+0.0665698325019955	3.906E-06
14	-1.000E+00	5.33188498285714E-06-2.11760594285714E-06i	0.40768420745186+0.0665677148960526i	1.856E-06
15	1.000E+00	-2.68585621333333E-06+4.90573508266667E-07i	0.407681521595647+0.0665682054695609	8.860E-07
16	-1.000E+00	0.0000013049863664+2.18426880000004E-08i	0.407682826582013+0.0665682273122489	4.247E-07
17	1.000E+00	-6.12055448847059E-07-1.33101158249412E-07i	0.407682214526565+0.0665680942110906	2.043E-07
18	-1.000E+00	0.000000276455519232+1.20658561564444E-07i	0.407682490982084+0.0665682148696522	9.861E-08
19	1.000E+00	-1.19521803277474E-07-8.33445783525053E-08i	0.407682371460281+0.0665681315250738	4.773E-08
20	-1.000E+00	4.8855121613312E-08+0.0000000509432460288i	0.407682420315402+0.0665681824683199	2.316E-08
21	1.000E+00	-1.84126059083581E-08-2.89115573102202E-08i	0.407682401902796+0.0665681535567626	1.126E-08

Figure 2.30: Output of an *Excel* program computing partial sum approximations of the complex variable logarithmic function at a user-specified evaluation point.

Figure 2.31: Contour plot of the real part of $\ln(z+1)$ for $x \in [0,1]$, $y \in [0,1]$. The contour plot was produced in *Excel* using the instructions given in Section 5.1.5.

Figure 2.32: Contour plot of the imaginary part of $\ln(z+1)$ for $x \in [0,1]$, $y \in [0,1]$. The contour plot was produced in *Excel* using the instructions given in Section 5.1.5.

2.5.5 Trigonometric functions

The two trigonometric functions of interest to us, $f(z) = \sin z$ and $f(z) = \cos z$, have definitions that can be derived using Euler's formula, as follows:

$$e^{iz} = \cos z + i \sin z \quad \text{and} \quad e^{-iz} = \cos z - i \sin z. \tag{2.53}$$

Note 2.5.5.1

In Equation (2.53), we have used the even and odd properties of cosine and sine, respectively:

$$\begin{aligned} &\text{Cosine is an even function:} \quad && \cos(-z) = \cos(z) \\ &\text{Sine is an odd function:} \quad && \sin(-z) = -\sin(z). \end{aligned} \tag{2.54}$$

Adding the expressions for e^{iz} and e^{-iz}, we obtain:

$$\begin{aligned} &e^{iz} + e^{-iz} = \cos z + i \sin z + (\cos z - i \sin z) = 2\cos(z) \\ &\implies \cos(z) = \frac{e^{iz} + e^{-iz}}{2}. \end{aligned} \tag{2.55}$$

Likewise, subtracting the expressions for e^{iz} and e^{-iz}, we obtain:

$$\begin{aligned} e^{iz} - e^{-iz} &= \cos z + i \sin z - (\cos z - i \sin z) = 2i \sin(z) \\ \implies \sin(z) &= \frac{e^{iz} - e^{-iz}}{2i}. \end{aligned} \tag{2.56}$$

Definition 2.5.5.1 Thus, we define the **complex cosine** and **complex sine** functions as follows. For $z \in \mathbb{C}$:

$$\begin{aligned} \cos(z) &= \frac{e^{iz} + e^{-iz}}{2} \\ \sin(z) &= \frac{e^{iz} - e^{-iz}}{2i} \end{aligned} \tag{2.57}$$

Note 2.5.5.2

The real variable cosine and sine functions have similar definitions as the complex variable cosine and sine functions. In particular, for $x \in \mathbb{R}$:

$$\begin{aligned} \cos(x) &= \frac{e^{ix} + e^{-ix}}{2} \\ \sin(x) &= \frac{e^{ix} - e^{-ix}}{2i}. \end{aligned} \tag{2.58}$$

Next, we compute the derivatives of both $\cos z$ and $\sin z$. Differentiating with respect to z, we find the derivatives to be as follows:

$$\begin{aligned} \frac{\mathrm{d}}{\mathrm{d}z} \cos z &= -\frac{e^{iz} - e^{-iz}}{2i} = -\sin z \\ \frac{\mathrm{d}}{\mathrm{d}z} \sin z &= \frac{e^{iz} + e^{-iz}}{2} = \cos z. \end{aligned} \tag{2.59}$$

With this in mind, the Taylor series expansion of both trigonometric functions will now be presented. The expansion for $\cos z$ where $|z| < \infty$ is:

$$\cos z = \sum_{n=0}^{\infty} (-1)^n \frac{z^{2n}}{(2n)!} = 1 - \frac{z^2}{2!} + \frac{z^4}{4!} - \dots \tag{2.60}$$

Similarly, the expansion of $\sin z$ where $|z| < \infty$ is:

$$\sin z = \sum_{n=0}^{\infty} (-1)^n \frac{z^{2n+1}}{(2n+1)!} = z - \frac{z^3}{3!} + \frac{z^5}{5!} - \ldots \tag{2.61}$$

Note 2.5.5.3

The Taylor series for the real variable cosine and sine functions are as follows. For $x \in \mathbb{R}$

$$\begin{aligned} \cos x &= \sum_{n=0}^{\infty} (-1)^n \frac{x^{2n}}{(2n)!} = 1 - \frac{x^2}{2!} + \frac{x^4}{4!} - \ldots \\ \sin x &= \sum_{n=0}^{\infty} (-1)^n \frac{x^{2n+1}}{(2n+1)!} = x - \frac{x^3}{3!} + \frac{x^5}{5!} - \ldots \end{aligned} \tag{2.62}$$

Cosine function

Tables 2.15–2.16 are provided to demonstrate the *Excel* syntax for computing partial sums of the Taylor series for the real variable cosine function. The provided code is for the Taylor series expanded about the point $x_0 = 0$ evaluated at the point $x = 3.5\pi$. Furthermore, the partial sum approximations $T_0, T_1, \ldots,$ T_{20} for the real variable cosine function are shown in Figure 2.33.

In Figure 2.33, the first column is used to define the value of the index n for the formula given in Equation (2.60), with a real variable input. The second and third columns are used to construct each term of Equation (2.60). The fourth column sums all previous terms together to produce the partial sum approximation of $\cos(3.5\pi)$. The fifth column computes an absolute error of the Taylor series partial sum approximation.

Notice how the absolute errors reported in the fifth column of Figure 2.33 demonstrate that the partial sums of the Taylor series appear to be converging.

Table 2.16 shows the second row of the *Excel* code for the Taylor series approximation of the real variable cosine function. This row is fully generalized and can be extended downwards to create a partial sum approximation using more terms. The use of more terms will result in a more accurate approximation of the real variable cosine function. Figure 2.33 displays the output of an *Excel* program computing partial sum approximations of the real variable cosine function at a user-specified evaluation point (in this case, $x = 3.5\pi$).

Excel Implementation 2.5.5.1

Cell	C2
Description	Real Evaluation Point
Code	=3.5*PI( )
Output	10.99557429

Table 2.15: This cell contains the real number at which the Taylor series partial sum will be evaluated.

Taylor Series Approximation of Real Cosine Function

ENTER Evaluation Point ----> x = 10.9955743

n	(2n)!	$(-1)^n x^{2n}/(2n)!$	Partial Sum Approximation	Absolute Error
0	1.000E+00	1.000E+00	1.000E+00	1.000E+00
1	2.000E+00	-6.045E+01	-5.945E+01	5.945E+01
2	2.400E+01	6.091E+02	5.496E+02	5.496E+02
3	7.200E+02	-2.455E+03	-1.905E+03	1.905E+03
4	4.032E+04	5.299E+03	3.394E+03	3.394E+03
5	3.629E+06	-7.119E+03	-3.725E+03	3.725E+03
6	4.790E+08	6.520E+03	2.796E+03	2.796E+03
7	8.718E+10	-4.332E+03	-1.536E+03	1.536E+03
8	2.092E+13	2.182E+03	6.464E+02	6.464E+02
9	6.402E+15	-8.621E+02	-2.157E+02	2.157E+02
10	2.433E+18	2.743E+02	5.858E+01	5.858E+01
11	1.124E+21	-7.178E+01	-1.321E+01	1.321E+01
12	6.204E+23	1.572E+01	2.515E+00	2.515E+00
13	4.033E+26	-2.924E+00	-4.098E-01	4.098E-01
14	3.049E+29	4.677E-01	5.785E-02	5.785E-02
15	2.653E+32	-6.499E-02	-7.143E-03	7.143E-03
16	2.631E+35	7.921E-03	7.782E-04	7.782E-04
17	2.952E+38	-8.536E-04	-7.537E-05	7.537E-05
18	3.720E+41	8.190E-05	6.533E-06	6.533E-06
19	5.230E+44	-7.043E-06	-5.098E-07	5.098E-07
20	8.159E+47	5.458E-07	3.601E-08	3.601E-08

Figure 2.33: Output of an *Excel* program computing partial sum approximations of the real variable cosine function at a user-specified evaluation point.

Tables 2.17–2.18 show the code and output for the complex variable equivalent of the implementation of the real variable cosine function given above. Like the real variable example, extending the code to include more terms in the partial sum will increase the accuracy of the approximation.

Table 2.18 shows the second row of the *Excel* code for the Taylor series approximation of the complex variable cosine function. Similar to the real variable example, this is a fully generalized row that can be extended downwards to create partial sum approximations with more terms. Figure 2.34 displays the *Excel* output for partial sum approximations of the complex variable cosine function at a user-specified evaluation point (in this case, $5 + 5i$).

Excel Implementation 2.5.5.2

Cell	A5	B5	C5	D5	E5
Description	n	$(2n)!$	$(-1)^n x^{2n}/(2n)!$	Partial Sum Approximation	Absolute Error
Code	0	=FACT(2*A5)	=POWER(-1,A5)	=SUM($C$5:C5)	=ABS(D5-COS
			*POWER		($C$2))
			($C$2,2*A5)/B5		
Output	0	1.000E+00	1.000E + 00	1.000E + 00	1.000E + 00

Cell	A6	B6	C6	D6	E6
Description	n	$(2n)!$	$(-1)^n x^{2n}/(2n)!$	Partial Sum Approximation	Absolute Error
Code	=A5+1	=FACT(2*A6)	=POWER(-1,A6)*	=SUM($C$5:C6)	=ABS(D6-COS
			POWER		($C$2))
			($C$2,2*A6)/B6		
Output	1	2.000E+00	−6.045E + 01	−5.945E + 01	5.945E + 01

Table 2.16: This is the second row of the *Excel* code for the Taylor series approximation of the real variable cosine function. This row is fully generalized and can be extended downwards to create a partial sum approximation using more terms.

Excel Implementation 2.5.5.3

Cell	C2
Description	Complex Evaluation Point
Code	=COMPLEX(5,5)
Output	$5 + 5i$

Table 2.17: This cell contains the complex value at which the Taylor series partial sum will be evaluated.

Taylor Series Approximation of Complex Cosine Function

ENTER Evaluation Point -----> z= 5+5i

n	(2n)!	$(-1)^n z^{2n}/(2n)!$	Partial Sum Approximation	Absolute Error
0	1.000E+00	1	1	7.393E+01
1	2.000E+00	-1.53143556863578E-15-25i	0.999999999999998-25i	9.822E+01
2	2.400E+01	-104.166666666667+1.27619630719648E-14i	-103.166666666667-25i	1.571E+02
3	7.200E+02	3.19049076799119E-14+173.611111111111i	-103.166666666667+148.611111111111i	1.464E+02
4	4.032E+04	155.009920634921-3.79820329522763E-14i	51.843253968254+148.611111111111i	8.335E+01
5	3.629E+06	-2.6376411772414E-14-86.1166225749559i	51.843253968254+62.4944885361551i	3.199E+01
6	4.790E+08	-32.6199327935439+1.198927807837E-14i	19.2233211747101+62.4944885361551i	8.851E+00
7	8.718E+10	3.8427173328109E-15+8.96151999822636i	19.2233211747101+71.4560085343815i	1.852E+00
8	2.092E+13	1.86698333296383-9.1493269828831E-16i	21.0903045076739+71.4560085343815i	3.034E-01
9	6.402E+15	-1.68186157773586E-16-0.3050626361052i	21.0903045076739+71.1509458982763i	3.998E-02
10	2.433E+18	-0.0401398205401579+2.45886195575419E-17i	21.0501646871338+71.1509458982763i	4.332E-03
11	1.124E+21	1.06439530586693E-17+0.00434413642209502i	21.0501646871338+71.1552900346984i	3.927E-04
12	6.204E+23	0.00039349061794339-2.89250653556628E-19i	21.0505581777517+71.1552900346984i	3.022E-05
13	4.033E+26	2.96634519797862E-20-0.0000302685090725683i	21.0505581777517+71.1552597661893i	1.999E-06
14	3.049E+29	-2.00188552067252E-06+1.71682486678911E-21i	21.0505561758662+71.1552597661893i	1.149E-07
15	2.653E+32	3.10087255790796E-22+1.15050891992673E-07i	21.0505561758662+71.1552598812402i	5.794E-09
16	2.631E+35	5.79893608834037E-09-5.68364607099525E-24i	21.0505561816651+71.1552598812402i	2.582E-10
17	2.952E+38	1.89933414943669E-25-2.58419611779875E-10i	21.0505561816651+71.1552598809818i	1.031E-11
18	3.720E+41	-1.02547464992014E-11+1.13072281460785E-26i	21.0505561816549+71.1552598809818i	4.101E-13
19	5.230E+44	1.07224383242175E-27+3.64678040512141E-13i	21.0505561816549+71.1552598809821i	1.407E-13
20	8.159E+47	1.16883987343635E-14-1.43200236497613E-29i	21.0505561816549+71.1552598809821i	1.407E-13

Figure 2.34: Output of an *Excel* program computing partial sum approximations of the complex variable cosine function at a user-specified evaluation point.

Excel Implementation 2.5.5.4

Cell	A5	B5	C5	D5	E5
Description	n	$(2n)!$	$(-1)^n z^{2n}/(2n)!$	Partial Sum Approximation	Absolute Error
Code	0	=FACT(2*A5)	=IMPRODUCT(POWER (-1,A5),IMDIV(IMPOWER ($C$2,2*A5),B5))	=IMSUM($C$5:C5)	=IMABS(IMSUB(D5, IMCOS($C$2)))
Output	0	1.000E+00	1	1	7.393E+01

Cell	A6	B6	C6	D6	E6
Description	n	$(2n)!$	$(-1)^n z^{2n}/(2n)!$	Partial Sum Approximation	Absolute Error
Code	=A5+1	=FACT(2*A6)	=IMPRODUCT(POWER (-1,A6),IMDIV(IMPOWER ($C$2,2*A6),B6))	=IMSUM($C$5:C6)	=IMABS(IMSUB(D6, IMCOS($C$2)))
Output	1	2.000E+00	$-1.531E-15-25i$	$0.999 - 25i$	9.822E+01

Table 2.18: This is the second row of the *Excel* code for the Taylor series approximation of the complex variable cosine function. This row is fully generalized and can be extended downwards to create a partial sum approximation using more terms.

Figure 2.35: Contour plot of the real part of $\cos(z)$ for $x \in [0,1]$, $y \in [0,1]$. The contour plot was produced in *Excel* using the instructions given in Section 5.1.5.

Figure 2.36: Contour plot of the imaginary part of $\cos(z)$ for $x \in [0,1]$, $y \in [0,1]$. The contour plot was produced in *Excel* using the instructions given in Section 5.1.5.

Sine function

Tables 2.19–2.20 are provided to demonstrate the *Excel* syntax for computing partial sums of the Taylor series for the real variable sine function. The provided code is for the Taylor series expanded about the point $x_0 = 0$ evaluated at the point $x = 3.5\pi$. Furthermore, the partial sum approximations $T_0, T_1, \ldots,$ T_{20} for the real variable sine function are shown in Figure 2.37.

In Figure 2.37, the first column is used to define the value of the index n for the formula given in Equation (2.61) with a real variable input. The second and third columns are used to construct each term of Equation (2.61). The fourth column sums all previous terms together to produce the partial sum approximation of $\sin(3.5\pi)$. The fifth column computes an absolute error of the Taylor series partial sum approximation.

Notice how the absolute errors reported in the fifth column of Figure 2.37 demonstrate that the partial sums of the Taylor series appear to be converging.

Table 2.20 shows the second row of the *Excel* code for the Taylor series approximation of the real variable exponential function. This row is fully generalized and can be extended downwards to create a partial sum approximation using more terms. The use of more terms will result in a more accurate approximation of the real variable exponential function. Figure 2.37 displays the output of an *Excel* program computing partial sum approximations of the real variable exponential function at a user-specified evaluation point (in this case, $x = 3.5\pi$).

Excel Implementation 2.5.5.5

Cell	C2
Description	Real Evaluation Point
Code	=3.5*PI( )
Output	10.99557429

Table 2.19: This cell contains the real number at which the Taylor series partial sum will be evaluated.

Excel Implementation 2.5.5.6

Cell	A5	B5	C5	D5	E5
Description	n	$(2n+1)!$	$(-1)^n x^{2n+1}/(2n+1)!$	Partial Sum Approximation	Absolute Error
Code	0	=FACT(2*A5+1)	=POWER(-1,A5)*POWER ($C$2,2*A5+1)/B5	=SUM($C$5:C5)	=ABS(D5-SIN($C$2))
Output	0	1.000E+00	1.100E+01	1.100E+01	1.200E+01

Cell	A6	B6	C6	D6	E6
Description	n	$(2n+1)!$	$(-1)^n x^{2n+1}/(2n+1)!$	Partial Sum Approximation	Absolute Error
Code	=A5+1	=FACT(2*A6+1)	=POWER(-1,A6)*POWER ($C$2,2*A6+1)/B6	=SUM($C$5:C6)	=ABS(D6-SIN($C$2))
Output	1	6.000E+00	−2.216E + 02	−2.106E + 02	2.096E + 02

Table 2.20: This is the second row of the *Excel* code for the Taylor series approximation of the real variable sine function. This row is fully generalized and can be extended downwards to create a partial sum approximation using more terms.

Tables 2.21–2.22 show the code and output for the complex variable equivalent of the implementation of the real variable sine function given above. Like the real variable example, extending the code to include more terms in the partial sum will increase the accuracy of the approximation.

Table 2.22 shows the second row of the *Excel* code for the Taylor series approximation of the complex variable sine function. Similar to the real variable example, this is a fully generalized row that can be extended downwards to create partial sum approximations with more terms. Figure 2.38 displays the *Excel* output for partial sum approximations of the complex variable sine function at a user-specified evaluation point (in this case, $5 + 5i$).

Taylor Series Approximation of Real Sine Function

ENTER Evaluation Point ----> x = 10.99557429

n	(2n+1)!	$(-1)^n x^{2n+1}/(2n+1)!$	Partial Sum Approximation	Absolute Error
0	1.000E+00	1.100E+01	1.100E+01	1.200E+01
1	6.000E+00	-2.216E+02	-2.106E+02	2.096E+02
2	1.200E+02	1.339E+03	1.129E+03	1.130E+03
3	5.040E+03	-3.856E+03	-2.727E+03	2.726E+03
4	3.629E+05	6.474E+03	3.748E+03	3.749E+03
5	3.992E+07	-7.116E+03	-3.369E+03	3.368E+03
6	6.227E+09	5.515E+03	2.147E+03	2.148E+03
7	1.308E+12	-3.175E+03	-1.029E+03	1.028E+03
8	3.557E+14	1.411E+03	3.827E+02	3.837E+02
9	1.216E+17	-4.989E+02	-1.162E+02	1.152E+02
10	5.109E+19	1.436E+02	2.744E+01	2.844E+01
11	2.585E+22	-3.432E+01	-6.881E+00	5.881E+00
12	1.551E+25	6.915E+00	3.430E-02	1.034E+00
13	1.089E+28	-1.191E+00	-1.157E+00	1.567E-01
14	8.842E+30	1.773E-01	-9.793E-01	2.066E-02
15	8.223E+33	-2.305E-02	-1.002E+00	2.394E-03
16	8.683E+36	2.639E-03	-9.998E-01	2.457E-04
17	1.033E+40	-2.682E-04	-1.000E+00	2.249E-05
18	1.376E+43	2.434E-05	-1.000E+00	1.849E-06
19	2.040E+46	-1.986E-06	-1.000E+00	1.372E-07
20	3.345E+49	1.464E-07	-1.000E+00	9.234E-09

Figure 2.37: Output of an *Excel* program computing partial sum approximations of the real variable sine function at a user-specified evaluation point.

Excel Implementation 2.5.5.7

Cell	C2
Description	Complex Evaluation Point
Code	=COMPLEX(5,5)
Output	$5 + 5i$

Table 2.21: This cell contains the complex number at which the Taylor series partial sum will be evaluated.

Excel Implementation 2.5.5.8

Cell	A5	B5	C5	D5	E5
Description	n	$(2n+1)!$	$(-1)^n z^{2n+1}/(2n+1)!$	Partial Sum Approximation	Absolute Error
Code	0	=FACT(2*A5+1)	=IMPRODUCT(POWER (-1,A5),IMDIV(IMPOWER ($C$2,2*A5+1),B5))	=IMSUM($C$5:C5)	=IMABS(IMSUB(D5, IMSIN($C$2)))
Output	0	1.000E+00	$5 + 5i$	$5 + 5i$	7.783E+01

Cell	A6	B6	C6	D6	E6
Description	n	$(2n+1)!$	$(-1)^n z^{2n+1}/(2n+1)!$	Partial Sum Approximation	Absolute Error
Code	=A5+1	=FACT(2*A6+1)	=IMPRODUCT(POWER (-1,A6),IMDIV(IMPOWER ($C$2,2*A6+1),B6))	=IMSUM($C$5:C6)	=IMABS(IMSUB(D6, IMSIN($C$2)))
Output	1	6.000E+00	$41.666 - 46.666i$	$46.666 - 36.666i$	1.312E+02

Table 2.22: This is the second row of the *Excel* code for the Taylor series approximation of the complex variable sine function. This row is fully generalized and can be extended downwards to create a partial sum approximation using more terms.

Taylor Series Approximation of Complex Sine Function

ENTER Evaluation Point ------> z =		5+5i		
n	(2n+1)!	$(-1)^n z^{2n+1}/(2n+1)!$	Partial Sum Approximation	Absolute Error
0	1.000E+00	5+5i	5+5i	7.783E+01
1	6.000E+00	41.6666666666667-41.6666666666667i	46.6666666666667-36.6666666666667i	1.312E+02
2	1.200E+02	-104.166666666667-104.166666666667i	-57.5000000000003-140.833333333334i	1.625E+02
3	5.040E+03	-124.007936507937+124.007936507937i	-181.507936507937-16.8253968253967i	1.167E+02
4	3.629E+05	86.1166225749559+86.1166225749559i	-95.3913139329814+69.2912257495592i	5.399E+01
5	3.992E+07	39.1439193522527-39.1439193522527i	-56.2473945807287+30.1473063973065i	1.747E+01
6	6.227E+09	-12.5461279975169-12.5461279975169i	-68.7935225782456+17.6011783997896i	4.183E+00
7	1.308E+12	-2.98717333274212+2.98717333274212i	-71.7806959109877+20.5883517325317i	7.714E-01
8	3.557E+14	0.549112744989359+0.549112744989359i	-71.2315831659984+21.1374644775211i	1.130E-01
9	1.216E+17	0.0802796410803157-0.080279641080316i	-71.151303524918+21.0571848364408i	1.347E-02
10	5.109E+19	-0.00955710012860903-0.00955710012860903i	-71.1608606250467+21.0476277363121i	1.332E-03
11	2.585E+22	-0.000944377483064132+0.000944377483064136i	-71.1618050025297+21.0485721137952i	1.111E-04
12	1.551E+25	0.0000786981235886777+0.0000786981235886777i	-71.1617263044061+21.0486508119188i	7.916E-06
13	1.089E+28	5.60527945788302E-06-5.60527945788304E-06i	-71.1617206991267+21.0486452066393i	4.876E-07
14	8.842E+30	-3.4515267597802E-07-3.4515267597802E-07i	-71.1617210442793+21.0486448614867i	2.622E-08
15	8.223E+33	-1.85565954826892E-08+1.85565954826893E-08i	-71.1617210628359+21.0486448800433i	1.242E-09
16	8.683E+36	8.78626680051573E-10+8.78626680051573E-10i	-71.1617210619573+21.0486448809219i	5.284E-11
17	1.033E+40	3.69170873971248E-11-3.6917087397125E-11i	-71.1617210619204+21.048644880885i	1.614E-12
18	1.376E+43	-1.38577655394613E-12-1.38577655394613E-12i	-71.1617210619218+21.0486448808836i	8.020E-13
19	2.040E+46	-4.67535949374537E-14+4.67535949374539E-14i	-71.1617210619218+21.0486448808836i	8.020E-13
20	3.345E+49	1.42541447980043E-15+1.42541447980042E-15i	-71.1617210619218+21.0486448808836i	8.020E-13

Figure 2.38: Output of an *Excel* program computing partial sum approximations of the complex variable sine function at a user-specified evaluation point.

Figure 2.39: Contour plot of the real part of $\sin(z)$ for $x \in [0,1]$, $y \in [0,1]$. The contour plot was produced in *Excel* using the instructions given in Section 5.1.5.

Figure 2.40: Contour plot of the imaginary part of $\sin(z)$ for $x \in [0,1]$, $y \in [0,1]$. The contour plot was produced in *Excel* using the instructions given in Section 5.1.5.

Conclusion

The methods demonstrated in this section can be used to approximate real and complex variable functions. Furthermore, we have demonstrated that the *Excel* implementation of Taylor series is relatively straightforward provided that a closed-form expression for the nth derivative of a function can be easily obtained. The ease with which Taylor series can be implemented makes them useful tools in function approximation.

Chapter 3

Contour integration

Bryce D. Wilkins
Matthew Fauerbach
John Lee

Contents

A major theme in complex analysis is the generalization of concepts in real analysis. Throughout this chapter, we will explore this theme particularly in the context of the complex line integral.

3.1 Open contour

In this section:

1. Fundamental theorem of complex line integrals.

2. Green's theorem.

3. Types of singularities.

Definition 3.1.0.1 Let Γ be a piecewise smooth curve and suppose $f(z)$ is continuous on Γ. We define the **complex integral of f over Γ** as

$$\int_{\Gamma} f(z)\,\mathrm{d}z = \int_{a}^{b} f\left(z(t)\right) z'(t)\,\mathrm{d}t, \quad a \le t \le b \tag{3.1}$$

where $z(t)$ is a parametrization of Γ.

Example 3.1.0.1

Consider the function $f(z) = z^3$. We wish to integrate this function over the straight line connecting the points $(0,0)$ and $(2,1)$, as depicted in Figure 3.1.

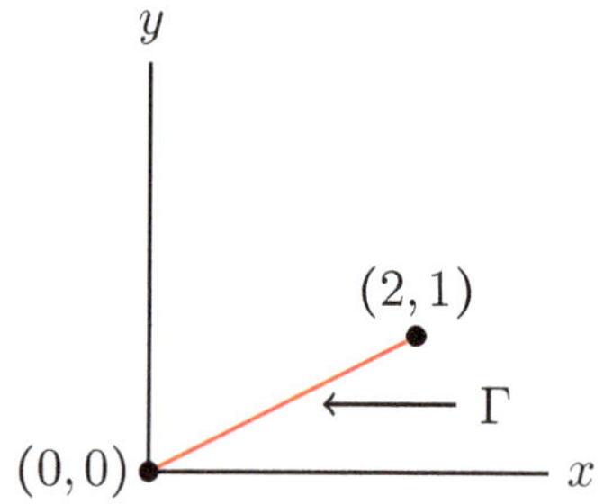

Figure 3.1: The contour is the straight line connecting the points $(0,0)$ and $(2,1)$.

The contour can be parametrized by the following

$$z(t) = 2t + it = (2+i)t \quad \text{where } 0 \le t \le 1. \tag{3.2}$$

Using this definition of $z(t)$, we determine $z'(t) = 2 + i$. Then, by Definition 3.1.0.1, the complex line integral is computed by

$$\begin{aligned}\int_{\Gamma} f(z)\,\mathrm{d}z &= \int_{a}^{b} f\left(z(t)\right) z'(t)\,\mathrm{d}t \\ &= (2+i)\int_{0}^{1} \left[(2+i)t\right]^3 \mathrm{d}t\end{aligned}$$

$$= (2+i)^4 \int_0^1 t^3\,\mathrm{d}t$$

$$= (-7+24i) \int_0^1 t^3\,\mathrm{d}t$$

$$\boxed{= -\frac{7}{4} + 6i}.$$

Excel Implementation 3.1.0.1

In the following demonstration, we assume cell A1 is initialized with the value of t at the left endpoint of the interval: $t = 0.005$ (we are using the midpoint rule, so the evaluation point is half-way in the interval [0,0.01]). The following table shows the commands in the second row, which can be generalized to all subsequent rows.

Column	A	B
Column Name	t	$z(t)$
Code	$= A1 + 0.01$	=IMPRODUCT(A2,COMPLEX(2,1))
Output	0.015	$0.03 + 0.015i$

Column	C	D
Column Name	$f(z(t))$	$z'(t)$
Code	=IMPOWER(B2,3)	=COMPLEX(2,1)
Output	$6.75001E - 06 + 0.000037125i$	$2 + i$

Column	E	F
Column Name	*dt*	Product
Code	= 0.01	=IMPRODUCT(C2,D2,E2)
Output	0.01	$-0.00000023625 + 0.00000081i$

Column	G
Column Name	Approximation
Code	=IMSUM(F:F)
Output (using 100 intervals)	$-1.7499 + 5.9997i$
Output (using 10,000 intervals)	$-1.7500 + 6.0000i$

Table 3.1: Demonstration of an open contour integration using the midpoint rule in *Excel*.

Recall that f has a real component and an imaginary component, which we denote $u(x,y)$ and $v(x,y)$, respectively:

$$f(z) = u(x,y) + iv(x,y). \tag{3.3}$$

Let $z(t) = x(t) + iy(t)$ be a parametrization for some piecewise smooth curve, Γ. Then, using Equation (3.1), we obtain

$$\begin{aligned}\int_\Gamma f(z)\,\mathrm{d}z &= \int_a^b f\left(z(t)\right) z'(t)\,\mathrm{d}t \\ &= \int_a^b \left[u\left(x(t),y(t)\right) + iv\left(x(t),y(t)\right)\right]\left[x'(t) + iy'(t)\right]\mathrm{d}t \\ &= \int_a^b \left[u\left(x(t),y(t)\right)x'(t) - v\left(x(t),y(t)\right)y'(t)\right]\mathrm{d}t \\ &\quad + i\int_a^b \left[v\left(x(t),y(t)\right)x'(t) + u\left(x(t),y(t)\right)y'(t)\right]\mathrm{d}t.\end{aligned} \tag{3.4}$$

Example 3.1.0.2

We consider the same problem as in Example 3.1.0.1. The contour can be parametrized by the following

$$\begin{aligned} z(t) &= 2t + it, \quad 0 \le t \le 1 \\ &= x(t) + iy(t) \quad \text{where } x(t) = 2t \text{ and } y(t) = t. \end{aligned} \tag{3.5}$$

Based on the definitions of $x(t)$ and $y(t)$, we have $x'(t) = 2$ and $y'(t) = 1$. Additionally, the real and imaginary parts of z^3 are given as, respectively,

$$\begin{aligned} u(x,y) &= \mathrm{Re}\left(z^3\right) = x^3 - 3xy^2 \\ v(x,y) &= \mathrm{Im}\left(z^3\right) = 3x^2y - y^3. \end{aligned} \tag{3.6}$$

Now, replacing x with $x(t)$ and y with $y(t)$, we obtain

$$\begin{aligned} u\left(x(t),y(t)\right) &= x(t)^3 - 3x(t)y(t)^2 = 8t^3 - 6t^3 = 2t^3 \\ v\left(x(t),y(t)\right) &= 3x(t)^2y(t) - y(t)^3 = 12t^3 - t^3 = 11t^3. \end{aligned} \tag{3.7}$$

Therefore, by Equation (3.4), the complex line integral is computed as

$$\int_\Gamma f(z)\,\mathrm{d}z = \int_0^1 \left[4t^3 - 11t^3\right]\mathrm{d}t + i\int_0^1 \left[22t^3 + 2t^3\right]\mathrm{d}t$$
$$= -7\int_0^1 t^3\,\mathrm{d}t + 24i\int_0^1 t^3\,\mathrm{d}t \tag{3.8}$$
$$\boxed{= -\frac{7}{4} + 6i}.$$

Fundamental theorem of complex line integrals

Theorem 3.1.0.2 *If $f(z) : \mathbb{C} \to \mathbb{C}$ is analytic in an open region, $\Omega \subseteq \mathbb{C}$, and Γ is a curve in Ω connecting the point $z_0 \in \Omega$ to $z_1 \in \Omega$, then:*

$$\int_\Gamma f'(z)\,\mathrm{d}z = f(z_1) - f(z_0). \tag{3.9}$$

Note 3.1.0.1

The fundamental theorem of complex line integrals is a logical extension of the fundamental theorem of calculus for real variables.

Path independence

Consider a function $f(z)$ defined in a region Ω. Let A and B be two points in Ω, and let $\Gamma_1 \subset \Omega$ and $\Gamma_2 \subset \Omega$ be two curves connecting points A and B. If the line integral $\int_\Gamma f(z)\,\mathrm{d}z$ is path independent, then:

$$\int_{\Gamma_1} f(z)\,\mathrm{d}z = \int_{\Gamma_2} f(z)\,\mathrm{d}z, \quad \forall \Gamma_1, \Gamma_2. \tag{3.10}$$

Note 3.1.0.2

If $f(z)$ is analytic in a simply connected domain, Ω, then $\int_\Gamma f(z)\,\mathrm{d}z$ is path independent for any path Γ contained in Ω.

Properties of complex integrals

Complex integrals have many of the same properties as real integrals because a complex function is a linear combination of two real functions, one of which is multiplied by the imaginary constant i, as in Equation (3.3).

Linearity: If $f(t) = u(t) + iv(t)$ and $g(t) = p(t) + iq(t)$ are both continuous on $a \leq t \leq b$.

$$\int_a^b (f(t) + g(t))dt = \int_a^b f(t)dt + \int_a^b g(t)dt.$$

Interchanging limits of integration: This property is essential for rules described in the remainder of the chapter.

$$\int_a^b f(t)dt = -\int_b^a f(t)dt.$$

Interchanging limit and integration operations: If f_n is a continuous function on a region Ω and f_n converges to f uniformly on Ω then f is continuous on Ω and for any contour Γ which lays in Ω

$$\lim_{n\to\infty} \int_\Gamma f_n(z)dz = \int_\Gamma f(z)dz.$$

Note that only uniform limits commute with contour integrals.

Types of isolated singularities

Poles: An analytic function, $f(z)$, has a pole of order m at z_0 if and only if m is the smallest positive integer for which $(z - z_0)^m f(z)$ is analytic at z_0.

Note 3.1.0.3
A pole of order 1 is known as a simple pole. Simple poles are the basis functions that will be used in the *Excel* implementation of the CVBEM program described in Part II of this book.

Note 3.1.0.4
Poles and zeros are related as follows: a pole of $f(z)$ is a zero of $\frac{1}{f(z)}$.

Removable singularities: A singularity at z_0 is removable if it is possible to redefine the function at the point z_0 such that the function becomes analytic in a neighborhood of z_0. Consider Riemann's Theorem of removable singularities. If z_0 is a singularity of $f(z)$ and $\lim_{z\to z_0}(z-z_0)f(z)=0$, then the singularity is removable.

Essential singularities: A singularity which is neither removable nor a pole is classified as an essential singularity.

Note 3.1.0.5

Picard's Great Theorem states an analytic function comes arbitrarily close to any complex value, except perhaps one, infinitely many times in any neighborhood around an essential singularity.

3.1.1 Examples of contour integration

The easiest class of curves Γ to consider is that of an open contour, which is a contour that does not have the same upper and lower endpoints.

Example 3.1.1.1

Evaluate $\int\limits_\Gamma z^2\,\mathrm{d}z$ where Γ is the line segment from $z_0=0$ to $z_1=1+i$.

We parameterize: $\Gamma(t)=t(1+i)$ for $0\le t\le 1$, then we find $\mathrm{d}\Gamma=(1+i)\,\mathrm{d}t$. It follows that the line integral is

$$\int_\Gamma z^2\,\mathrm{d}z=\int_0^1 t^2(1+i)^2(1+i)\,\mathrm{d}t=\left(\frac{2i(1+i)}{3}\right). \tag{3.11}$$

Alternatively we could use the fundamental theorem of complex line integrals. Consider that $z^2=\frac{\mathrm{d}}{\mathrm{d}z}\frac{z^3}{3}$. Therefore,

$$\int_\Gamma z^2\,\mathrm{d}z=\int_\Gamma\left(\frac{\mathrm{d}}{\mathrm{d}z}\frac{z^3}{3}\right)\mathrm{d}z=\left(\frac{z^3}{3}\right)\Bigg|_{z=0}^{z=1+i}=\left(\frac{2i(1+i)}{3}\right). \tag{3.12}$$

Note 3.1.1.1

The Estimation Lemma

The Estimation Lemma, also known as the ML Inequality or ML Lemma, states that if $f(z)$ is integrable along a curve Γ with a finite length L and a value M, such that $|f(z)| \leq M$ on Γ, it then follows that

$$\left| \int_\Gamma f(z)dz \right| \leq ML. \quad (3.13)$$

The proof is based on our definition of contour integrals.

Proof.

$$\int_\Gamma f(z)dz = \lim_{n\to\infty} \sum_{k=1}^{n} f(\zeta_k)\Delta z_k$$

The Triangle Inequality for complex numbers states $|z_1 + z_2| \leq |z_1| + |z_2|$. Therefore, $\left|\sum_{k=1}^{n} f(\zeta_k)\Delta z_k\right| \leq \sum_{k=1}^{n} |f(\zeta_k)||\Delta z_k|$. Recall in our statement of the problem $|f(z)| \leq M$.

$$\left| \sum_{k=1}^{n} f(\zeta_k)\Delta z_k \right| \leq M \sum_{k=1}^{n} |\Delta z_k| \quad (3.14)$$

It is clear that $\sum_{k=1}^{n} |\Delta z_k| \leq L$ as L is the total length of Γ:

$$\left| \sum_{k=1}^{n} f(\zeta_k)\Delta z_k \right| \leq ML. \quad (3.15)$$

This can be rewritten as

$$\left| \int_\Gamma f(z)\,\mathrm{d}z \right| \leq \int_\Gamma |f(z)||\mathrm{d}z|. \quad (3.16)$$

□

A numerical approach to Example 3.1.1.1

Consider the basic definition of the Riemann integral: the limit of the Riemann sums as the number of components tends to infinity. We carry this same idea over to complex line integrals. Using a prescribed number of subdivisions, we can obtain increasingly accurate estimations of the value of the integral. Formally we see this as $S_n = \sum_{k=1}^{n} f(\zeta_k)\Delta z_k$ where z_k is an arc of arbitrary size we have broken the contour up into. $f(\zeta_k)$ is the value of the function at a particular point along the contour somewhere on the arc z_k and Δz_k is the interval between two adjacent arcs along the contour. As in normal Riemann sums, ζ_k can be any point within the arc z_k although we typically choose the midpoint. This is shown in the figure below.

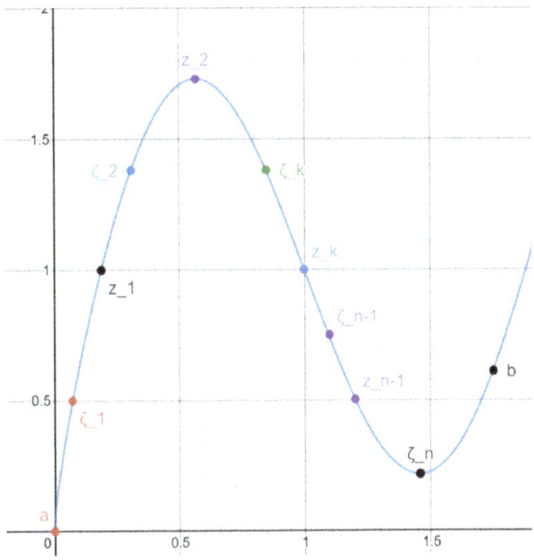

Figure 3.2: Depiction of a standard Riemann sum.

One benefit of using Riemann sums is that it is not necessary to evaluate the integral analytically. Moreover, a sufficiently accurate approximation of the integral can usually be obtained by shrinking the width of the intervals between evaluation points. However, obtaining a high level of precision using this technique would be tedious to do by hand. For this reason, it is convenient to use *Excel* to perform these computations.

Simple open contour *Excel* computation

We will demonstrate an efficient way to approximate the line integral from Example 3.1.1.1 using the program *Excel*. We parameterize our curve in an identical fashion, setting $\Gamma(t) = t + it$. Then we substitute for z^2 and obtain:

$$\int_\Gamma z^2 \, \mathrm{d}z = \int_0^1 t^2(1+i)^2(1+i)\, \mathrm{d}t\,. \tag{3.17}$$

Let $f(t)$ denote the integrand. The next step is to discretize the interval of integration, $[0,1]$. Suppose there are n intervals with endpoints given by $t_0, t_1, \ldots, t_{n-1}, t_n$. Define $\Delta t_k = t_k - t_{k-1}$ for $k = 1, 2, \ldots, n$. Additionally, within each interval $[t_{k-1}, t_k]$, we need to determine the location at which we will evaluate the integrand. There are various techniques for this step including the left (or right) endpoint rule, the midpoint rule, and the trapezoid rule, among others. Let $t_k^* \in [t_{k-1}, t_k]$ denote the evaluation point selected by one of these rules. Then, the approximation of the integral, denoted I, is carried out as follows:

$$I \approx \sum_{k=1}^{n} f(t_k^*)\Delta t_k.$$

In general, as the number of intervals increases, so will the accuracy of the approximation. In the example that follows, we will use 100 uniformly-spaced intervals. Therefore $\Delta t_k = \frac{1}{100} = 0.01$ for all $k = 1, 2, \ldots, 100$.

Excel Implementation 3.1.1.1

In the following demonstration, we assume cell A1 is initialized with the value of t at the left endpoint of the interval: $t = 0$. The following table shows the commands in the second row, which can be generalized to all subsequent rows.

Column	A	B
Column Name	t	t^2
Code	=A1 + 0.01	=POWER(A2,2)
Output	0.01	0.0001

Column	C	D
Column Name	$(i+i)$	$(1+i)^2$
Code	=COMPLEX(1,1)	=IMPOWER(C2,2)
Output	$1+i$	$0+2i$

Column	E	F
Column Name	Product	Product ∗ dt
Code	=IMPRODUCT(B2,C2,D2)	=IMPRODUCT(E2,0.01)
Output	$-0.0002+0.0002i$	$-0.000002+0.000002i$

Column	G
Column Name	Approximation
Code	=IMSUM(F:F)
Output (using 100 intervals)	$-0.6767+0.6767i$
Output (using 10,000 intervals)	$-0.6667+0.6667i$

Table 3.2: Demonstration of an open contour integration in *Excel*.

Notice that we give the final 100th output for each column. This is because in order to apply our integration across 100 increments, we must conduct *each* operation at its respective increment. This is performed in *Excel* by grabbing the corner of the first cell in a column and dragging it down to the 100th cell. Doing this for each column will yield our desired integration. Finally, we attain the solution through summing the contents of the final product, which we can check analytically to be $\frac{-2}{3} + \frac{2}{3}i$.

Example 3.1.1.2

We now consider a line integral over a move complicated contour. Suppose the contour of interest takes the shape of the cosine function, as replicated in Figure 3.3. In this case, the contour is discretized to visualize the process of averaging two points more easily. Consider each point a node for the use in the previous algorithm and rework Example 3.1.1.1 along the path from −360 degrees to 360 degrees with the background function $f(z) = z^2$. In particular, note doing this computationally requires only a few subtle changes to the program rather than doing the problem from the beginning.

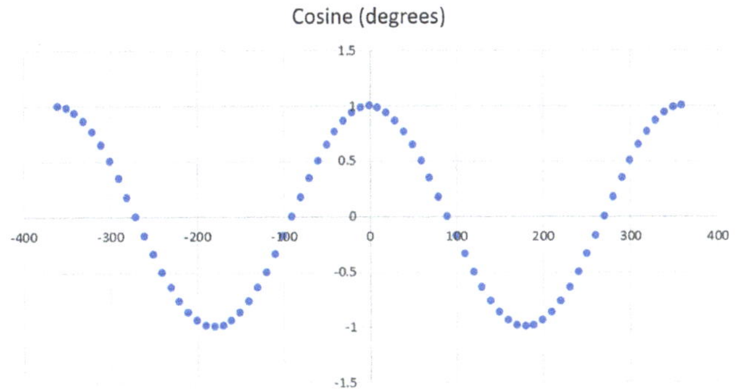

Figure 3.3: Depiction of the discretization of the cosine contour of interest. Line integration over more complicated contours such as this can be more difficult than computing the line integral over a straight-line contour.

Consider a slight modification to the problem where instead of terminating the straight line at $1 + i$, it connects back to the origin such as in the below graph.

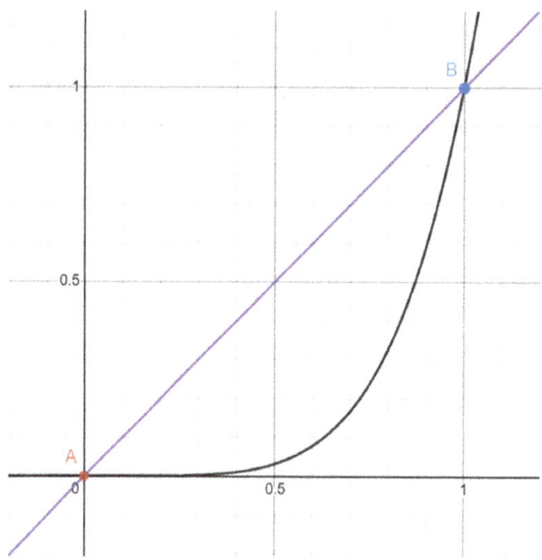

Figure 3.4: Depiction of a simple closed contour. When the background function is analytic, the contour integral over a simple, closed curve is always 0, as described in Section 3.2.

Consider how this slight modification has completely changed the problem, at least analytically. We could take the integral of both halves of the problem then add them together to find the solution, but there is an easier method that is available to us because the background function is analytic and the contour of interest is simple and closed.

3.2 Simple, closed contour

In this section:

1. Cauchy–Goursat Theorem.
2. Cauchy–Riemann equations.

This section makes the natural transition to closed curves, specifically simple closed contours. If we traverse the path in the direction such that the bounded component is always to our left, then we say that we traverse the contour in the positive direction. If the path is near circular, this would correspond to going around it in the counterclockwise direction. If the bounded component is on the right as the path is traversed, we travel in the negative direction. If not stated otherwise, all simple closed curves are traversed in the positive direction.

Cauchy–Goursat Theorem

If a function $f(z)$ is analytic at all points interior to and on the boundary of a simple closed curve Γ, then

$$\int_{\Gamma} f(z)\,\mathrm{d}z = 0.$$

Note 3.2.0.1

Cauchy's proof of the theorem, demonstrated below, requires the additional hypothesis that f' is continuous, although Goursat later provided a proof which removed this restriction. Recall that $\int_{\Gamma} f(z)\,\mathrm{d}z = \int_{\Gamma} u\,\mathrm{d}x - v\,\mathrm{d}y + i\int_{\Gamma} v\,\mathrm{d}x + u\,\mathrm{d}y$. By applying Green's Theorem to both the real and imaginary parts of the right side, we obtain the following:

$$\int_{\Gamma} u\,\mathrm{d}x - v\,\mathrm{d}y = \iint_{R} -\frac{\partial v}{\partial x} - \frac{\partial u}{\partial y}\,\mathrm{d}x\,\mathrm{d}y$$

$$\int_{\Gamma} v\,\mathrm{d}x + u\,\mathrm{d}y = \iint_{R} \frac{\partial u}{\partial x} - \frac{\partial v}{\partial y}\,\mathrm{d}x\,\mathrm{d}y$$

$$\int_{\Gamma} f(z)\,\mathrm{d}z = \iint_{R} -\frac{\partial v}{\partial x} - \frac{\partial u}{\partial y}\,\mathrm{d}x\,\mathrm{d}y + i\iint_{R} \frac{\partial u}{\partial x} - \frac{\partial v}{\partial y}\,\mathrm{d}x\,\mathrm{d}y\,.$$

As previously stated, the function $f(z)$ is analytic at all points in and on Γ, so we can apply the Cauchy–Riemann equations.

$$\frac{\partial u}{\partial x} = \frac{\partial v}{\partial y} \quad \text{and} \quad \frac{\partial u}{\partial y} = -\frac{\partial v}{\partial x}. \tag{3.18}$$

The integrands simplify to:

$$\int_{\Gamma} f(z)\,\mathrm{d}z = \iint_{R} 0\ \mathrm{d}x\,\mathrm{d}y + i \iint_{R} 0\ \mathrm{d}x\,\mathrm{d}y = 0. \tag{3.19}$$

Example 3.2.0.1

Integrate the function $f(z) = z^2$ along the square with vertices (0,0), (0,1), (1,0), and (1,1) depicted below.

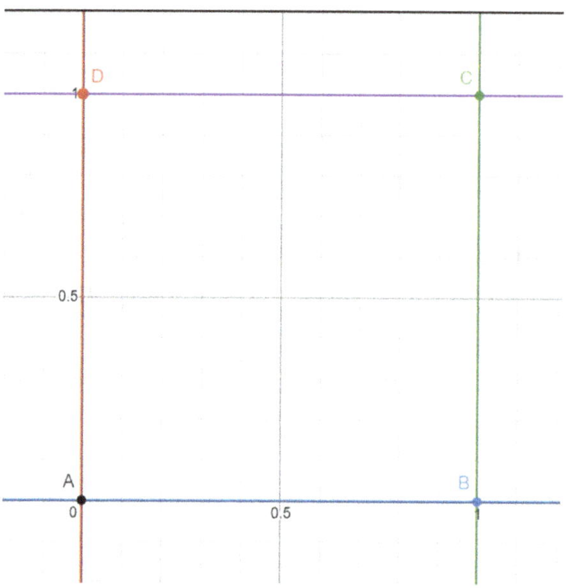

Figure 3.5: Problem domain for demonstration of the Cauchy–Goursat Theorem. The problem domain is $\Omega = [0,1] \times [0,1]$.

By the Cauchy–Goursat Theorem, we know the answer should be 0 because z^2 is analytic everywhere in and on our region Ω. Assume a positive direction of travel, and it is clear that by breaking the contour integral up into 4 separate line integrals, they will all cancel. The integral over $\Gamma = \partial\Omega$ is computed over the following four line segments.

$$\int_{\Gamma} z^2\,\mathrm{d}z = \int_{A}^{B} z^2\,\mathrm{d}z + \int_{B}^{C} z^2\,\mathrm{d}z + \int_{C}^{D} z^2\,\mathrm{d}z + \int_{D}^{A} z^2\,\mathrm{d}z\,. \tag{3.20}$$

However, the following integrals are equivalent.

$$\int_{A}^{B} z^2\,\mathrm{d}z = -\int_{C}^{D} z^2\,\mathrm{d}z \qquad \int_{B}^{C} z^2\,\mathrm{d}z = -\int_{D}^{A} z^2\,\mathrm{d}z\,. \tag{3.21}$$

Therefore,

$$\oint_{\Gamma} z^2\,\mathrm{d}z = \int_{A}^{B} z^2\,\mathrm{d}z + \int_{B}^{C} z^2\,\mathrm{d}z - \int_{A}^{B} z^2\,\mathrm{d}z - \int_{B}^{C} z^2\,\mathrm{d}z = 0. \tag{3.22}$$

We can demonstrate this process analytically with $A = (0,0)$, $B = (a, 0)$, $C = (a,b)$, and $D = (0,b)$. Let $I = \int_A^B z^2\,\mathrm{d}z$. Thus,

$$\begin{aligned}\int_A^B z^2\,\mathrm{d}z &= \int_0^a x^2\,\mathrm{d}x + \int_0^b (a+iy)^2 i\,\mathrm{d}y \\ &\quad + \int_a^0 (x+ib)^2\,\mathrm{d}x + \int_b^0 (iy)^2 i\,\mathrm{d}y\,.\end{aligned} \tag{3.23}$$

Continuing by hand and integrating, we can see that this simplifies to the following:

$$\frac{a^3}{3} + \left(\frac{2}{3} + \frac{2i}{3}\right) - 2ia - (1-i)a^2 + 2ia + (1-i)a^2 - \frac{a^3}{3} - \frac{2}{3} - \frac{2i}{3} = 0. \tag{3.24}$$

In this case, while evaluating these integrals exactly may have been tedious, they were still do-able by hand. However, for most problems of interest, this is not usually the case. This is the reason why numerical integration (quadrature) is such an important skill.

In the program given in the *Excel* Implementation 3.2.0.1, we move along the problem boundary in increments of 0.1. To increase accuracy, we can always make this increment smaller, though 0.1 is sufficient in this case to show an integration value of 0. For Columns A and B, we show the first ten cells' worth of code and the next ten since they differ. Initially, we are moving along the bottom edge of the domain, so x is changing by 0.1 with each increment while yi remains fixed at 0. However, once we get to the 11th cell, now yi is increasing by 0.1 while x is fixed at 0 along the right edge.

The process continues in a similar fashion for the top and left edges of Ω. We should also take care to notice that our last three rows of our table, indicating the final sum code and final sum value, are separate cells which in our case are placed in Column F, but are not part of the product. Finally, note that in this *Excel* computation we keep our integral in terms of z rather than converting to t. While we could convert to t seeing as we are in fact parameterizing each side with either $x = t$ or $y = t$, the same effect is achieved by leaving the integral in terms of z and changing the code of Columns A and B as explained.

Excel Implementation 3.2.0.1

Consider the simple closed contour Ω from our Example 3.2.0.1. As was stated before, we know from the Cauchy–Goursat Theorem that integrating z^2 on $\partial\Omega$ will give us a value of 0 since z^2 is analytic everywhere and on Ω. Once again, *Excel* is useful for numerically approximating these integrals, particularly when they become more complex and are difficult to compute by hand. To check our methods, we compute the integral across this contour since we know we can verify our calculations directly.

Similar to before, we seek to find a way to break down our contour into discrete parts such that we can sum them up to approximate an integral. Consider how for Ω we can parameterize one side at a time. That is, along the bottom edge y is fixed at 0 while $x = t$. Along the right edge x is fixed at 1 while $y = t$, and so on, as we move counter-clockwise from the origin. While it is possible to analytically calculate the integrals we set up in 1.3.2 using *Excel*, we are most interested in numerical capabilities from the program. The following steps show one way to compute the solution numerically.

Column	A	B	C
Column Name	x	yi	z
Row 2 Code	=A1 + 0.1	=0	=COMPLEX(A1,B1)
Row 11 Code	=1	=B10 + 0.1	COMPLEX(A1,B1)
Row 21 Code	=A20 – 0.1	=1	COMPLEX(A1,B1)
Row 31 Code	=0	=B30 – 0.1	COMPLEX(A1,B1)
Final Output	0	0	0

Column	D	E	F
Column Name	z^2	dz	Product
Code	=IMPOWER (COMPLEX (A1,B1),2)	=IMSUB(D2,D1)	=IMPRODUCT(IMDIV(IMSUB(C2, C1),2),E1)
Final Output	0	$-0.1i$	$-0.0005i$
Final Sum Code			=IMSUM(F1:F40)
Final Sum Value			0

Table 3.3: Demonstration of a simple closed contour integration.

Example 3.2.0.2

While the shape of the contour can make the problem more complicated, the biggest complication comes from having singularities in the background function of the area of interest. Consider the background function $f(z) = \frac{e^z}{(z-1)^3}$ on a circle of radius 3, centered at 0.

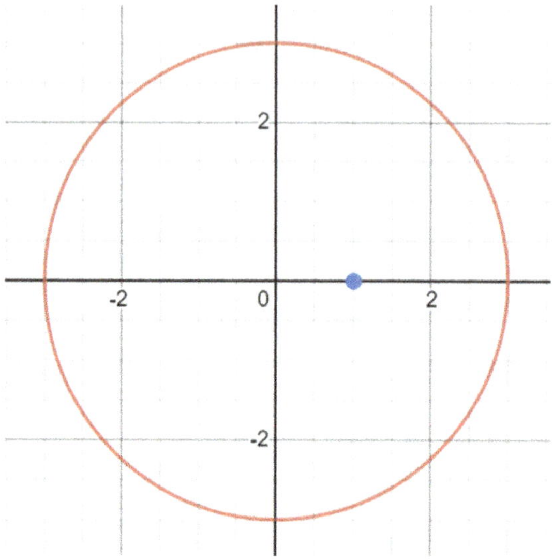

Figure 3.6: Depiction of a contour that encloses a singularity of the background function. In this case, the background function is $f(z) = \frac{e^z}{(z-1)^3}$, and there is a singularity located at $z = 1$.

At first, this may seem like an easy modification of the last problem. However, the singularity located at $z = 1$ is a complication and means we cannot simply apply the Cauchy–Goursat Theorem. To solve this problem we will need a theorem further explained in the next section.

3.3 The Cauchy integral formula

In this section:

1. The Cauchy integral formula.
2. The Cauchy integral formula for higher derivatives.
3. Computation of integral on a square boundary.

The Cauchy integral theorem is one of the most powerful tools in complex analysis.

If $f(z)$ is analytic in a domain Ω and Γ is a simple closed curve in Ω whose interior is also contained in Ω, and z_0 is any point interior to Γ, then the value of $f(z_0)$ is given by:

$$f(z_0) = \frac{1}{2\pi i} \int_\Gamma \frac{f(z)\,\mathrm{d}z}{z - z_0} \tag{3.25}$$

Note 3.3.0.1

Lemma: If a function is analytic in a region bounded by 2 curves C_0 and C_1, such that C_1 is entirely enclosed by C_0, then $\oint_{C_0} f(z)\,\mathrm{d}z = \oint_{C_1} f(z)\,\mathrm{d}z$.

We begin by noticing that the function $\frac{f(z)}{z-z_0}$ is analytic everywhere except at $z = z_0$.

Using the above lemma, we can rewrite the integral of Equation (3.25) as:

$$\oint_\Gamma \frac{f(z)}{z - z_0}\,\mathrm{d}z = \oint_\rho \frac{f(z)}{z - z_0}\,\mathrm{d}z\,, \tag{3.26}$$

where ρ is a circle of radius ε centered at z_0. Furthermore,

$$\begin{aligned} |z - z_0| &= \varepsilon \\ z - z_0 &= \varepsilon e^{i\theta}, \quad 0 \le \theta < 2\pi \\ z &= z_0 + \varepsilon e^{i\theta} \\ \mathrm{d}z &= i\varepsilon e^{i\theta}\,\mathrm{d}\theta\,. \end{aligned} \tag{3.27}$$

Substituting these values into Equation (3.26), we obtain:

$$\begin{aligned} \oint_\rho \frac{f(z)}{z - z_0}\,\mathrm{d}z &= \int_0^{2\pi} \frac{f(z_0 + \varepsilon e^{i\theta})}{z_0 + \varepsilon e^{i\theta} - z_0} i\varepsilon e^{i\theta}\,\mathrm{d}\theta \\ &= i\int_0^{2\pi} f(z_0 + \varepsilon e^{i\theta})\,\mathrm{d}\theta\,. \end{aligned} \tag{3.28}$$

Taking advantage of the continuity of $f(z)$, we take the limit as ε approaches 0.

$$\begin{aligned} \lim_{\varepsilon\to 0} i\int_0^{2\pi} f(z_0 + \varepsilon e^{i\theta})\,\mathrm{d}\theta &= \oint_\rho \frac{f(z)}{z - z_0}\,\mathrm{d}z \\ i\int_0^{2\pi} f(z_0)\,\mathrm{d}\theta &= \oint_\rho \frac{f(z)}{z - z_0}\,\mathrm{d}z \\ 2\pi i f(z_0) &= \oint_\rho \frac{f(z)}{z - z_0}\,\mathrm{d}z \\ f(z_0) &= \frac{1}{2\pi i}\int_\Gamma \frac{f(z)\,\mathrm{d}z}{z - z_0} \end{aligned} \tag{3.29}$$

Example 3.3.0.1

Consider the contour, Γ, depicted in Figure 3.7. We have a square with its bottom left corner at the origin, and the side-lengths of the square are 1. This is the same situation as in the previous *Excel* computation. We will use the same function as before, $f(z) = z^2$. Let $z_0 = \frac{1}{2} + \frac{1}{2}i$. This pole represents the center of the square Ω. It follows by the Cauchy integral formula that

$$f\left(\frac{1}{2} + \frac{1}{2}i\right) = \frac{1}{2\pi i} \int_{\Gamma} \frac{z^2 \, dz}{z - \left(\frac{1}{2} + \frac{1}{2}i\right)}.$$

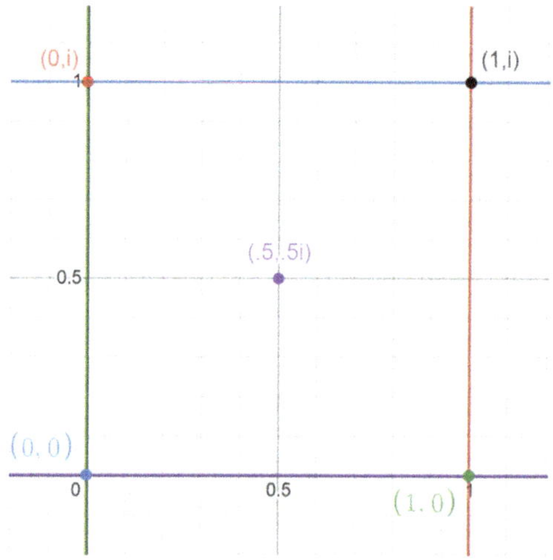

Figure 3.7: The problem domain is the unit square with the bottom-left corner located at the origin. The evaluation point is $\frac{1}{2} + \frac{1}{2}i$.

Substituting $\frac{1}{2} + \frac{1}{2}i$ into z^2 yields $0.5i$. However, often we will only know the boundary conditions of our contour, making it useful to calculate the integral on the right side of our equation. We will do this computation in *Excel*.

Excel Implementation 3.3.0.1

Since the contour we use for this integral is the same as the contour from our closed contour computation example, much of the *Excel* code remains the same. However, since we also must use $z - z_0$ and $\frac{1}{2\pi i}$, we must perform additional calculations. The specific steps are shown in Table 3.4.

Column	A	B	C
Column Name	x	yi	z
Row 2 Code	=A1 + 0.1	=0	=COMPLEX (A1,B1)
Row 11 Code	=1	=B10 + 0.1	COMPLEX (A1,B1)
Row 21 Code	=A20 − 0.1	=1	COMPLEX (A1,B1)
Row 31 Code	=0	=B30 − 0.1	COMPLEX (A1,B1)
Final Output	0	0	0

Column	D	E	F
Column Name	z^2	dz	$z - z_0$
Code	=IMPOWER (C,2)	=IMSUB (C2,C1)	=IMSUB(C1, COMPLEX (0.5,0.5))
Final Output	0	$-0.1i$	$-0.5 - 0.5i$

Column	G	H	I
Column Name	$z^2/z - z_0$	Product	Constant
Code	=IMDIV (D1,F1)	=IMPRODUCT (IMSUM (G2,G1), IMDIV(E1,2))	=IMDIV(1, COMPLEX (0,2*3.1415))
Final Output	0	-0.00048- 0.00061i	$-0.16i$
Final Sum Code			=IMSUM(H2:41)
Final Sum Value			−3.13
Final Value Code			=IMPRODUCT (I3,I1)
Final Value			$0.4998i$

Table 3.4: *Excel* computation of a Cauchy integral.

It is important to draw attention to a few key aspects of this *Excel* program. First, we are using a step size of 0.1 rather than a step size of 0.01. This change is made simply for the reader's convenience in generating the *Excel* table, as it reduces the amount of cells used. While decreasing step size will increase accuracy, our end solution ends up coming within 0.001 of our analytic solution, which is adequate for demonstration purposes. Next, we should be sure that we are not adding an empty space in the last cell when we are calculating an average with the trapezoid rule. This can be mitigated by offsetting Columns E and H by 1 cell from the other columns. Thirdly, our last four rows of our table represent values in the I column that are individually calculated, with I1 being the constant outside the integral and the next two being the total sum and the total sum times the constant respectively. Lastly, we must multiply the final sum by the constant $\frac{1}{2\pi i}$, which explains our Column I and the last two Final Value rows of our program.

Example 3.3.0.2

Now consider the circle $|z| = 3$, but with a different background function: $f(z) = \frac{e^{z^2}}{z-2}$. Evaluate $\int_\Gamma \frac{e^{z^2}}{z-2}\, dz$.

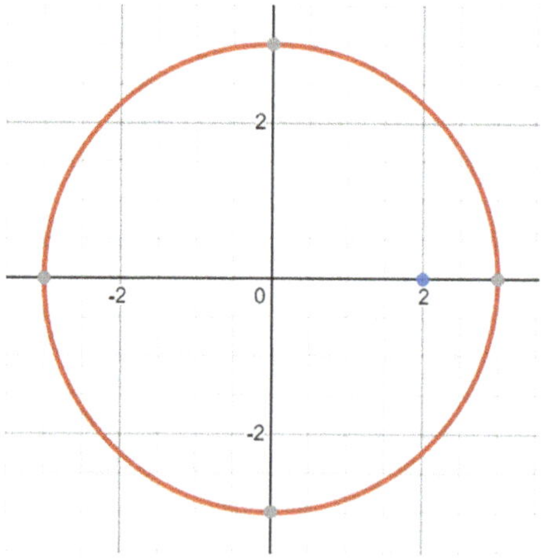

Figure 3.8: The contour of interest is a circle of radius 3, centered at the origin. There is a singularity in the background function at $z = 2$. Hence, the background function is not analytic inside the contour of interest.

Note the pole is inside the circle, so we use the Cauchy integral formula.

$$f(z_0) = \frac{1}{2\pi i} \int_{\Gamma} \frac{f(z)\,\mathrm{d}z}{z - z_0}$$

With $f(z) = e^{z^2}$ and $z_0 = 2$ we find the following:

$$2\pi i(e^{2^2}) = \int_{\Gamma} \frac{e^{z^2}}{z-2}\,\mathrm{d}z$$
$$2\pi i(e^4) = \int_{\Gamma} \frac{e^{z^2}}{z-2}\,\mathrm{d}z\,.$$

Example 3.3.0.3

Consider the same background function but with a shifted curve as shifted below.

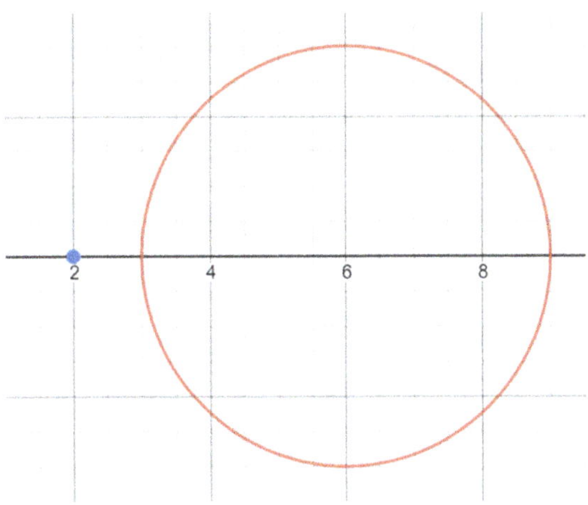

Figure 3.9: The contour of interest is a circle of radius 3, centered at (6, 0). There is a singularity in the background function at $z = 2$; however, the singularity is no longer located within the contour of interest, and the background function is analytic within the contour of interest.

How does the pole being outside of the curve effect the integral? With the pole outside of the curve we can use Cauchy–Goursat Theorem since the function is analytic within and on Γ. Hence,

$$\int_\Gamma \frac{e^{z^2}}{z-2}\,\mathrm{d}z = 0. \tag{3.30}$$

Cauchy integral formula for higher derivatives

This is perhaps the most important of the corollaries we draw from the Cauchy integral formula. If $f(z)$ is analytic on an open set Ω which contains a positively oriented circle Γ and its interior and z_0 is interior to Γ then f is n times complex differential at z_0 and

$$f^{(n)}(z_0) = \frac{n!}{2\pi i}\int\limits_\Gamma \frac{f(z)\,\mathrm{d}z}{(z-z_0)^{n+1}}$$

Using the normal definition $0! = 1$, the case of $n = 0$ yields our standard formula. This constitutes the base case for the following proof:

Proof. We have already proven our base case of $n = 0$ above. Recall the limit definition of the derivative is as follows:

$$\begin{aligned} f'(z_0) &= \lim_{z\to z_0} \frac{f(z)-f(z_0)}{z-z_0} \\ &= \lim_{z\to z_0} \frac{1}{z-z_0}|f(z)-f(z_0)| \\ &= \lim_{z\to z_0} \frac{1}{z-z_0}\left(\frac{1}{2\pi i}\int_\Gamma \frac{f(w)}{w-z}\,\mathrm{d}w - \frac{1}{2\pi i}\int_\Gamma \frac{f(w)}{w-z_0}\,\mathrm{d}w\right). \end{aligned}$$

Applying the linearity of integrals we simplify the following:

$$\begin{aligned} f'(z_0) &= \lim_{z\to z_0} \frac{1}{z-z_0}\frac{1}{2\pi i}\int_\Gamma \left(\frac{f(w)}{w-z} - \frac{f(w)}{w-z_0}\right)\mathrm{d}w \\ &= \lim_{z\to z_0} \frac{1}{z-z_0}\frac{1}{2\pi i}\int_\Gamma \frac{f(w)(w-z_0)-f(w)(w-z)}{(w-z)(w-z_0)}\,\mathrm{d}w \\ &= \lim_{z\to z_0} \frac{1}{z-z_0}\frac{1}{2\pi i}\int_\Gamma \frac{wf(w)-z_0f(w)-wf(w)+zf(w)}{(w-z)(w-z_0)}\,\mathrm{d}w \end{aligned}$$

$$= \lim_{z \to z_0} \frac{1}{z - z_0} \frac{1}{2\pi i} \int_\Gamma \frac{f(w)(z - z_0)}{(w - z)(w - z_0)} \, \mathrm{d}w$$
$$= \lim_{z \to z_0} \frac{1}{2\pi i} \int_\Gamma \frac{f(w)}{(w - z)(w - z_0)} \, \mathrm{d}w$$

By uniform convergence we interchange the limit and the integral.

$$f'(z_0) = \frac{1}{2\pi i} \int_\Gamma \lim_{z \to z_0} \frac{f(w)}{(w - z)(w - z_0)} \, \mathrm{d}w$$
$$= \frac{1}{2\pi i} \int_\Gamma \frac{f(w)}{(w - z_0)^2} \, \mathrm{d}w$$

A simple change of variables yields the required result below.

$$f'(z_0) = \frac{1}{2\pi i} \int_\Gamma \frac{f(z)}{(z - z_0)^2} \, \mathrm{d}z$$

□

Every other case follows by the principle of mathematical induction.

Note 3.3.0.2

Of special note is that Cauchy's integral formula only applies in cases where the isolated singularities are either removable or poles. If any of the singularities are essential, then to compute the contour integral we refer to Cauchy's residue theorem.

Example 3.3.0.4

We now seek to demonstrate an example of the first derivative of the Cauchy integral formula. For the sake of continuity, we return to our example of the unit square Ω with the contour defined as $\partial\Omega$. Again, the background function is z^2, and the evaluation point is $z_0 = 0.5 + 0.5i$. If we compute the integral analytically, we obtain:

$$f'(0.5 + 0.5i) = \frac{1}{2\pi i} \int_\Omega \frac{z^2}{\left[z - (0.5 + 0.5i)\right]^2} \, \mathrm{d}z$$

We can see that the left side of the equation is equivalent to $f'(0.5 + 0.5i) = 2(0.5 + 0.5i) = 1 + i$. Numerically, we know that our right side should evaluate to $1 + i$. We will solve this problem computationally.

Excel Implementation 3.3.0.2

Numerical computation of the Cauchy integral formula's derivatives is an important part of understanding the properties of analytic functions. While $f(z)$ is a simple enough calculation with our example, this will not always be the case. For this reason, a familiarity with *Excel* numerical approximation of these integrals (the right side of our equation) is necessary. We need only adjust our $z - z_0$ term from our last calculation to $(z - z_0)^2$ in this case to find the first derivative. The calculation steps are shown in Table 3.5.

Column	A	B	C
Column Name	x	yi	z
Row 2 Code	=A1+0.1	=0	=COMPLEX (A1,B1)
Row 11 Code	=1	=B10+0.1	COMPLEX (A1,B1)
Row 21 Code	=A20-0.1	=1	COMPLEX (A1,B1)
Row 31 Code	=0	=B30-0.1	COMPLEX (A1,B1)
Final Output	0	0	0

Column	D	E	F
Column Name	z^2	dz	$(z - z_0)^2$
Code	=IMPOWER(C,2)	=IMSUB(C2,C1)	=IMPOWER(IMSUB(C1, COMPLEX (0.5,0.5)),2)
Final Output	0	$-0.1i$	$0.5i$

Column	G	H	I
Column Name	$z^2/(z-z_0)^2$	Product	Constant
Code	=IMDIV(D1,F1)	=IMPRODUCT (IMSUM (G2,G1), IMDIV(E1,2))	=IMDIV (1,COMPLEX (0,2*3.1415))
Final Output	0	0.0012+ 0.0003i	−0.16i
Final Sum Code			=IMSUM(H2:41)
Final Sum Value			−6.2699+ 6.2699i
Final Value Code			=IMPRODUCT (I4,I1)
Final Value			0.9979+ 0.9979i

Table 3.5: Calculation steps for a Cauchy Integral Formula First Derivative.

Our final value of $0.9979 + 0.9979i$ is within 0.001 of our analytic solution of $1 + 1i$. This *Excel* program is useful in that it shows only minor changes are needed to get the first derivative evaluated at z_0 from our approximation of the standard value. All that was necessary was changing the $z - z_0$ to $(z - z_0)^2$. It should be noted that in higher orders of derivatives, the constant term would also change by the order of a factorial, in accordance with our equation.

Applications of the Cauchy integral formula

The Cauchy integral formula allows us to extend the Leibniz rule for interchanging derivatives and integrals to complex contour integrals. It is also trivial to show that an analytic function has continuous derivatives of every order, something which we have taken for granted in previous proofs. Further, by applying the ML Lemma, we are able to derive the Cauchy inequality, Liouville's theorem, and the Fundamental Theorem of Algebra.

Theorem 3.3.0.1 *The Cauchy Inequality. If $f(z)$ is analytic inside and on a circle Γ centered at z_0 with radius r and $|f(z)| \leq M$, then*

$$\left|f^{(n)}(z_0)\right| \leq M\frac{n!}{r^n}. \tag{3.31}$$

Theorem 3.3.0.2 *Liouville's Theorem. Any entire analytic function that is bounded is constant. This is a consequence of the Cauchy Inequality which we proved using the Cauchy integral formula. As we bring the radius to infinity then the rate of change goes to infinity and the function must be constant. Apart from being an excellent party fact, Liouville's Theorem provides one of many proofs for the fundamental theorem of algebra which is in itself quite astounding. The connection between algebra and complex analysis is tenuous, but nonetheless the proof exists and is quite elegant.*

Theorem 3.3.0.3 *The Fundamental Theorem of Algebra. Every polynomial equation* $P(z) = a_0 + a_1 n + a_2 n^2 + \ldots + a_n z^n$ *where* $n \geq 1$ *and* $a_n \neq 0$ *has at least one root. The proof from Liouville's Theorem works by contradiction. If there are no roots then* $\frac{1}{P(z)}$ *is analytic and* $\frac{1}{|P(z)|}$ *is bounded. This implies by Liouville's theorem that* $P(z)$ *is constant, a contradiction to its previous definition of a polynomial.*

Example 3.3.0.5

With this new information, we return to an example from the previous section. The circle of radius 3 has a background function $f(z) = \frac{e^z}{(z-1)^3}$. The integral on the curve Γ is given as $\int_\Gamma \frac{e^z}{(z-1)^3}\,\mathrm{d}z$.

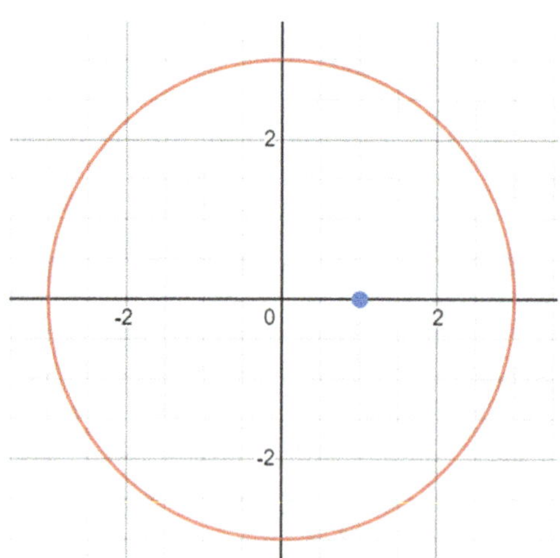

Figure 3.10: The contour of interest is a circle of radius 3, centered at the origin. The background function has a singularity at $z = 1$.

Recall the Cauchy integral formula for derivatives:

$$f^{(n)}(z_0) = \left(\frac{n!}{2\pi i}\right) \int_{\Gamma} \left(\frac{f(z)}{(z - z_0)^{n+1}}\right) \mathrm{d}z\,. \tag{3.32}$$

Equation (3.32) is reduced as follows for $n = 2$, $f(z) = e^z$ and $z_0 = 1$:

$$e = \frac{1}{\pi i} \int_{\Gamma} \frac{e^z}{(z-1)^3}\, \mathrm{d}z\,. \tag{3.33}$$

Thus,

$$\int_{\Gamma} \frac{e^z}{(z-1)^3}\, \mathrm{d}z = ei\pi. \tag{3.34}$$

Part II

The complex variable boundary element method and *Excel*

Chapter 4

Foundations of the CVBEM

Bryce D. Wilkins
Theodore V. Hromadka II

Contents

The Complex Variable Boundary Element Method is a numerical technique for developing approximate solutions to boundary value problems of the Laplace equation. These types of problems are common in potential theory. In this chapter, we examine foundational material related to the CVBEM.

4.1 Analytic solutions of Laplace's equation and related PDEs

In this section:

1. Partial differential equations and boundary value problems.
2. Analytic solutions to Laplace's equation in two-dimensional circular and rectangular domains.
3. Laplace's equation in Cartesian and polar coordinates.

Definition 4.1.0.1 Let $u(x, y)$ denote a two-dimensional real variable function with continuous second-order partial derivatives. **Laplace's equation in two spatial dimensions** is defined as follows:

$$\nabla \cdot \nabla u = \nabla^2 u = \frac{\partial^2 u}{\partial x^2} + \frac{\partial^2 u}{\partial y^2} = 0. \tag{4.1}$$

Any two-dimensional real variable function satisfying Laplace's equation is known as a **harmonic function**.

Example 4.1.0.1

Consider the function $u(x, y) = x^2 - y^2$. We will verify that this function is a harmonic function by determining if u satisfies Equation (4.1).

$$\frac{\partial^2 u}{\partial x^2} + \frac{\partial^2 u}{\partial y^2} = 2 - 2 = 0 \tag{4.2}$$

Since u satisfies Laplace's equation, u is said to be a harmonic function.

Definition 4.1.0.2 Let Ω denote a simply-connected problem domain with boundary $\partial\Omega$. A **boundary value problem** is a differential equation along with additional constraints specified on $\partial\Omega$, which are known as **boundary conditions**. A solution to a boundary value problem is a function satisfying the differential equation in Ω and the boundary conditions on $\partial\Omega$.

To illustrate the idea of a boundary value problem, we will find the analytic solution to a boundary value problem of the Laplace equation. For many boundary value problems, it is not possible to determine an analytic solution. However, for problems with simple domain geometries, such as either a rectangle or a disk, it can be possible to obtain an exact solution. We will consider one of these instances in Example 4.1.0.2.

Example 4.1.0.2

Consider the following boundary value problem. In this example problem, we will consider a disk problem domain. Because of the geometry of the problem domain, it is most convenient to use polar coordinates.

$$\begin{array}{lll} \text{Domain:} & 0 \leq r < 3, & \text{where } r^2 = x^2 + y^2 \\ \text{PDE:} & \nabla^2 u = 0 & \\ \text{Boundary Conditions:} & u(3,\theta) = \cos^2(3\theta), & 0 \leq \theta \leq 2\pi \end{array} \tag{4.3}$$

The boundary in this problem is a circle centered at the origin of radius 3. The boundary conditions for this problem are given by $u(3,\theta) = \cos^2(3\theta)$.

Recall the half angle formula:

$$\cos^2\left(\frac{\theta}{2}\right) = \frac{1+\cos\theta}{2}. \tag{4.4}$$

We will use Equation (4.4) to obtain a more convenient expression for the problem boundary conditions:

$$u(3,\theta) = \cos^2(3\theta) = \cos^2\left(\frac{6\theta}{2}\right) = \frac{1+\cos 6\theta}{2} = \frac{1}{2} + \frac{1}{2}\cos 6\theta. \tag{4.5}$$

Now, consider the complex polynomials given by:

$$z^n, \quad n \geq 0 \tag{4.6}$$

Complex polynomials are analytic throughout the complex plane.

Using the transformation to polar coordinates, $z = re^{i\theta}$, Equation (4.6) is equivalent to:

$$z^n = \left(re^{i\theta}\right)^n = r^n e^{in\theta} = r^n \cos(n\theta) + ir^n \sin(n\theta). \tag{4.7}$$

We have applied Euler's identity to obtain the final result in Equation (4.7).

Since the polynomials of Equation (4.6) are analytic throughout the complex plane, both the real and imaginary parts of those polynomials are harmonic functions. Thus, the following functions are harmonic:

$$\begin{aligned} \mathrm{Re}(z^n) &= r^n \cos(n\theta), \qquad n \geq 0 \\ \mathrm{Im}(z^n) &= r^n \sin(n\theta), \qquad n \geq 0. \end{aligned} \tag{4.8}$$

Moreover, since Laplace's equation is linear, its solutions obey the principle of superposition. Therefore,

$$u(r,\theta) = \sum_{n=0}^{\infty} a_n r^n \cos(n\theta) + \sum_{n=0}^{\infty} b_n r^n \sin(n\theta) \tag{4.9}$$

is a solution to Laplace's equation, where the coefficients $a_n \in \mathbb{R}$ and $b_n \in \mathbb{R}$ will be determined based on the boundary conditions.

In order to match the boundary conditions, we are particularly interested in Equation (4.9) when all of the coefficients except for a_0 and a_6 are zero:

$$u(r,\theta) = a_0 + a_6 r^6 \cos(6\theta) \tag{4.10}$$

We are able to determine the necessary coefficients of this linear combination by matching the form of Equation (4.10) to the boundary conditions given in Equation (4.5). Hence, we conclude

$$a_0 = \frac{1}{2} \quad \text{and} \quad a_6 r^6 = a_6 (3)^6 = \frac{1}{2} \implies a_6 = \frac{1}{2 \cdot 3^6}. \tag{4.11}$$

Therefore, the solution to this boundary value problem is

$$\boxed{u(r,\theta) = \frac{1}{2} + \frac{r^6 \cos(6\theta)}{2 \cdot 3^6}}. \tag{4.12}$$

In Example 4.1.0.2, it was possible to obtain an analytic solution to the boundary value problem because of the simplicity of the domain geometry, in this case, a disk. For boundary value problems with more complicated domain geometries, it is not usually possible to determine an exact solution expressed in a closed form. Rather, when the domain geometry or the boundary conditions are more complicated, it is usually necessary to employ a numerical technique to obtain an approximation of the exact solution.

Before discussing numerical techniques, let us verify that the solution we obtained in Equation (4.12) actually satisfies Laplace's equation and the boundary conditions. Since we solved the boundary value problem in Example 4.1.0.2 using polar coordinates, it is useful to define Laplace's equation in polar coordinates.

Definition 4.1.0.3 Let $u(r, \theta)$ denote a two-dimensional real variable function with continuous second-order partial derivatives. **Laplace's equation in polar coordinates** is defined as follows:

$$\nabla \cdot \nabla u = \nabla^2 u = \frac{\partial^2 u}{\partial r^2} + \frac{1}{r}\frac{\partial u}{\partial r} + \frac{1}{r^2}\frac{\partial^2 u}{\partial \theta^2} = 0 \tag{4.13}$$

The polar form of Laplace's equation is derived from the Cartesian form as follows: The polar conversions are as follows:

$$x = r\cos\theta \quad \text{and} \quad y = r\sin\theta. \tag{4.14}$$

Important partial derivatives that we will use later are as follows:

$$\begin{aligned} \frac{\partial x}{\partial r} &= \cos\theta & \frac{\partial y}{\partial r} &= \sin\theta \\ \frac{\partial x}{\partial \theta} &= -r\sin\theta & \frac{\partial y}{\partial \theta} &= r\cos\theta. \end{aligned} \tag{4.15}$$

Therefore:

$$\frac{\partial u}{\partial r} = \frac{\partial u}{\partial x}\frac{\partial x}{\partial r} + \frac{\partial u}{\partial y}\frac{\partial y}{\partial r} = \frac{\partial u}{\partial x}\cos\theta + \frac{\partial u}{\partial y}\sin\theta. \tag{4.16}$$

Furthermore:

$$\frac{\partial^2 u}{\partial r^2} = \frac{\partial}{\partial r}\frac{\partial u}{\partial r} = \frac{\partial}{\partial x}\left[\frac{\partial u}{\partial x}\cos\theta + \frac{\partial u}{\partial y}\sin\theta\right]\frac{\partial x}{\partial r} + \frac{\partial}{\partial y}\left[\frac{\partial u}{\partial x}\cos\theta + \frac{\partial u}{\partial y}\sin\theta\right]\frac{\partial y}{\partial r}$$
$$= \frac{\partial}{\partial x}\left[\frac{\partial u}{\partial x}\cos\theta + \frac{\partial u}{\partial y}\sin\theta\right]\cos\theta + \frac{\partial}{\partial y}\left[\frac{\partial u}{\partial x}\cos\theta + \frac{\partial u}{\partial y}\sin\theta\right]\sin\theta$$
$$= \left[\frac{\partial^2 u}{\partial x^2}\cos\theta + \frac{\partial^2 u}{\partial x\partial y}\sin\theta\right]\cos\theta + \left[\frac{\partial^2 u}{\partial x\partial y}\cos\theta + \frac{\partial^2 u}{\partial y^2}\sin\theta\right]\sin\theta$$
$$= \frac{\partial^2 u}{\partial x^2}\cos^2\theta + 2\frac{\partial^2 u}{\partial x\partial y}\sin\theta\cos\theta + \frac{\partial^2 u}{\partial y^2}\sin^2\theta. \tag{4.17}$$

Now, we compute $\frac{\partial u}{\partial \theta}$:

$$\frac{\partial u}{\partial \theta} = \frac{\partial u}{\partial x}\frac{\partial x}{\partial \theta} + \frac{\partial u}{\partial y}\frac{\partial y}{\partial \theta} = -r\sin\theta\frac{\partial u}{\partial x} + r\cos\theta\frac{\partial u}{\partial y}. \tag{4.18}$$

Furthermore:

$$\frac{\partial^2 u}{\partial \theta^2} = \frac{\partial}{\partial \theta}\frac{\partial u}{\partial \theta} = -r\sin\theta\frac{\partial}{\partial \theta}\frac{\partial u}{\partial x} - r\cos\theta\frac{\partial u}{\partial x} + r\cos\theta\frac{\partial}{\partial \theta}\frac{\partial u}{\partial y} - r\sin\theta\frac{\partial u}{\partial y}$$
$$= -r\left(\cos\theta\frac{\partial u}{\partial x} + \sin\theta\frac{\partial u}{\partial y}\right) + \frac{\partial^2 u}{\partial x\partial \theta}(-r\sin\theta) + \frac{\partial^2 u}{\partial y\partial \theta}(r\cos\theta)$$
$$= -r\left(\cos\theta\frac{\partial u}{\partial x} + \sin\theta\frac{\partial u}{\partial y}\right) + \frac{\partial}{\partial x}\frac{\partial u}{\partial \theta}(-r\sin\theta) + \frac{\partial}{\partial y}\frac{\partial u}{\partial \theta}(r\cos\theta) \tag{4.19}$$
$$= -r\left(\cos\theta\frac{\partial u}{\partial x} + \sin\theta\frac{\partial u}{\partial y}\right) + \frac{\partial}{\partial x}\left[-r\sin\theta\frac{\partial u}{\partial x} + r\cos\theta\frac{\partial u}{\partial y}\right](-r\sin\theta)$$
$$+ \frac{\partial}{\partial y}\left[-r\sin\theta\frac{\partial u}{\partial x} + r\cos\theta\frac{\partial u}{\partial y}\right](r\cos\theta)$$
$$= -r\left(\cos\theta\frac{\partial u}{\partial x} + \sin\theta\frac{\partial u}{\partial y}\right) + \left[-r\sin\theta\frac{\partial^2 u}{\partial x^2} + r\cos\theta\frac{\partial^2 u}{\partial x\partial y}\right](-r\sin\theta)$$
$$+ \left[-r\sin\theta\frac{\partial^2 u}{\partial x\partial y} + r\cos\theta\frac{\partial^2 u}{\partial y^2}\right](r\cos\theta)$$
$$= -r\left(\cos\theta\frac{\partial u}{\partial x} + \sin\theta\frac{\partial u}{\partial y}\right) + \left[r^2\sin^2\theta\frac{\partial^2 u}{\partial x^2} - r^2\cos\theta\sin\theta\frac{\partial^2 u}{\partial x\partial y}\right]$$
$$+ \left[-r^2\cos\theta\sin\theta\frac{\partial^2 u}{\partial x\partial y} + r^2\cos^2\theta\frac{\partial^2 u}{\partial y^2}\right]$$

$$
\begin{aligned}
&= -r\left(\cos\theta\frac{\partial u}{\partial x} + \sin\theta\frac{\partial u}{\partial y}\right) + r^2\cos^2\theta\frac{\partial^2 u}{\partial y^2} - 2r^2\cos\theta\sin\theta\frac{\partial^2 u}{\partial x\partial y}\\
&\quad + r^2\sin^2\theta\frac{\partial^2 u}{\partial x^2}\\
&= -r\left(\cos\theta\frac{\partial u}{\partial x} + \sin\theta\frac{\partial u}{\partial y}\right)\\
&\quad + r^2\left(\cos^2\theta\frac{\partial^2 u}{\partial y^2} - 2\cos\theta\sin\theta\frac{\partial^2 u}{\partial x\partial y} + \sin^2\theta\frac{\partial^2 u}{\partial x^2}\right).
\end{aligned}
$$

Finally, substituting these partial derivatives into Laplace's equation, we have:

$$
\begin{aligned}
\Delta u &= \frac{1}{r}\frac{\partial}{\partial r}\left(r\frac{\partial u}{\partial r}\right) + \frac{1}{r^2}\frac{\partial^2 u}{\partial\theta^2}\\
&= \frac{1}{r}\frac{\partial}{\partial r}\left(r\frac{\partial u}{\partial r}\right) + \frac{1}{r^2}\Bigg[-r\left(\cos\theta\frac{\partial u}{\partial x} + \sin\theta\frac{\partial u}{\partial y}\right)\\
&\quad + r^2\left(\cos^2\theta\frac{\partial^2 u}{\partial y^2} - 2\cos\theta\sin\theta\frac{\partial^2 u}{\partial x\partial y} + \sin^2\theta\frac{\partial^2 u}{\partial x^2}\right)\Bigg]\\
&= \frac{1}{r}\frac{\partial}{\partial r}\left(r\frac{\partial u}{\partial r}\right) - \frac{1}{r}\left(\cos\theta\frac{\partial u}{\partial x} + \sin\theta\frac{\partial u}{\partial y}\right)\\
&\quad + \left(\cos^2\theta\frac{\partial^2 u}{\partial y^2} - 2\cos\theta\sin\theta\frac{\partial^2 u}{\partial x\partial y} + \sin^2\theta\frac{\partial^2 u}{\partial x^2}\right)\\
&= \frac{\partial^2 u}{\partial r^2} + \frac{1}{r}\frac{\partial u}{\partial r} - \frac{1}{r}\left(\cos\theta\frac{\partial u}{\partial x} + \sin\theta\frac{\partial u}{\partial y}\right)\\
&\quad + \left(\cos^2\theta\frac{\partial^2 u}{\partial y^2} - 2\cos\theta\sin\theta\frac{\partial^2 u}{\partial x\partial y} + \sin^2\theta\frac{\partial^2 u}{\partial x^2}\right)
\end{aligned}
$$

$$
\begin{aligned}
&= \frac{\partial^2 u}{\partial r^2} + \underbrace{\frac{1}{r}\left(\frac{\partial u}{\partial x}\cos\theta + \frac{\partial u}{\partial y}\sin\theta\right) - \frac{1}{r}\left(\cos\theta\frac{\partial u}{\partial x} + \sin\theta\frac{\partial u}{\partial y}\right)}_{0}\\
&\quad + \left(\cos^2\theta\frac{\partial^2 u}{\partial y^2} - 2\cos\theta\sin\theta\frac{\partial^2 u}{\partial x\partial y} + \sin^2\theta\frac{\partial^2 u}{\partial x^2}\right)
\end{aligned}
\tag{4.20}
$$

$$
= \frac{\partial^2 u}{\partial x^2}\cos^2\theta + 2\frac{\partial^2 u}{\partial x\partial y}\sin\theta\cos\theta + \frac{\partial^2 u}{\partial y^2}\sin^2\theta + \cos^2\theta\frac{\partial^2 u}{\partial y^2}
$$

$$
\begin{aligned}
&- 2\cos\theta\sin\theta\frac{\partial^2 u}{\partial x\partial y} + \sin^2\theta\frac{\partial^2 u}{\partial x^2} \\
&= \frac{\partial^2 u}{\partial x^2}\underbrace{\left(\cos^2\theta + \sin^2\theta\right)}_{1} + \frac{\partial^2 u}{\partial x\partial y}\underbrace{\sin\theta\cos\theta(2-2)}_{0} \\
&\quad + \frac{\partial^2 u}{\partial y^2}\underbrace{\left(\cos^2\theta + \sin^2\theta\right)}_{1} \\
&= \frac{\partial^2 u}{\partial x^2} + \frac{\partial^2 u}{\partial y^2}.
\end{aligned}
$$

Thus, we have shown:

$$
\Delta u = \frac{\partial^2 u}{\partial x^2} + \frac{\partial^2 u}{\partial y^2} = \frac{1}{r}\frac{\partial}{\partial r}\left(r\frac{\partial u}{\partial r}\right) + \frac{1}{r^2}\frac{\partial^2 u}{\partial \theta^2} = \frac{\partial^2 u}{\partial x^2} + \frac{\partial^2 u}{\partial y^2}. \tag{4.21}
$$

Example 4.1.0.3

Using Definition 4.1.0.3, we can verify that the solution obtained in Equation (4.12) satisfies Laplace's equation. First, we will calculate the various partial derivatives that will be necessary to employ Definition 4.1.0.3.

$$
\begin{aligned}
\frac{\partial u}{\partial r} &= \frac{r^5\cos(6\theta)}{243} \\
\frac{\partial^2 u}{\partial r^2} &= \frac{5r^4\cos(6\theta)}{243} \\
\frac{\partial^2 u}{\partial \theta^2} &= -\frac{6r^6\cos(6\theta)}{243}
\end{aligned} \tag{4.22}
$$

Therefore, it follows that

$$
\begin{aligned}
\frac{\partial^2 u}{\partial r^2} + \frac{1}{r}\frac{\partial u}{\partial r} + \frac{1}{r^2}\frac{\partial^2 u}{\partial \theta^2} &= \frac{5r^4\cos(6\theta)}{243} + \frac{1}{r}\frac{r^5\cos(6\theta)}{243} - \frac{1}{r^2}\frac{6r^6\cos(6\theta)}{243} \\
&= \frac{5r^4\cos(6\theta)}{243} + \frac{r^4\cos(6\theta)}{243} - \frac{6r^4\cos(6\theta)}{243} \\
&= \frac{6r^4\cos(6\theta)}{243} - \frac{6r^4\cos(6\theta)}{243} \\
&= 0.
\end{aligned}
$$

Thus, we conclude that Equation (4.12) satisfies Laplace's equation. To verify that Equation (4.12) satisfies the given boundary conditions, observe:

$$u(3,\theta) = \frac{1}{2} + \frac{3^6\cos(6\theta)}{2\cdot 3^6} = \frac{1}{2} + \frac{\cos(6\theta)}{2} = \cos^2(3\theta)\,. \tag{4.23}$$

Since Equation (4.12) satisfies Laplace's equation and the given boundary conditions, we conclude that Equation (4.12) is a solution of the boundary value problem given in Equation (4.3).

Example 4.1.0.4

Consider the problem domain $\Omega = [0,L]\times[0,H]$. Suppose u satisfies Laplace's equation in Ω with boundary conditions given by:

$$\begin{aligned} \frac{\partial u}{\partial x}(0,y) = 4y \quad &\text{and} \quad \frac{\partial u}{\partial x}(L,y) = 0 \\ u(x,0) = 0 \quad &\text{and} \quad u(x,H) = 0. \end{aligned} \tag{4.24}$$

We assume the standard product solution: $u(x,y) = h(x)\phi(y)$. Substituting this into Laplace's equation, we obtain:

$$h''(x)\phi(y) + h(x)\phi''(y) = 0 \implies h''(x)\phi(y) = -h(x)\phi''(y). \tag{4.25}$$

Next, divide by $h(x)\phi(y)$ to obtain:

$$\frac{h''(x)}{h(x)} = -\frac{\phi''(y)}{\phi(y)}. \tag{4.26}$$

For a function of x to always be equal to a function of y, both functions must be constant. Therefore, we introduce the separation constant, λ, as follows:

$$\begin{aligned} h''(x) &= \lambda h(x) \quad &\text{with} \quad h'(0) = 0 \quad &\text{and} \quad h'(L) = 4y \\ \phi''(y) &= -\lambda\phi(y) \quad &\text{with} \quad \phi(0) = 0 \quad &\text{and} \quad \phi(H) = 0. \end{aligned} \tag{4.27}$$

We chose the separation constant to be λ instead of $-\lambda$ because this choice implies $\phi(y)$ oscillates while $h(x)$ is composed of exponentials. This situation seems more realistic given the boundary conditions.

However, we will verify this assumption by checking each of the possible eigenvalue cases.

In this case, the y-dependent problem is the boundary value problem because it has two homogeneous boundary conditions. We will use this problem to determine the allowable eigenvalues.

We assume $\phi(y)$ has exponential solutions of the form e^{ry}. When substituted into the ODE, this leads to:

$$r^2 e^{ry} = -\lambda e^{ry}. \tag{4.28}$$

This implies $r = \pm\sqrt{-\lambda}$. Hence, the exponential solution is of the form: $\phi(y) = e^{\pm\sqrt{-\lambda}y}$. Now, we consider three cases.

Case 1: λ > 0. In this case, $-\lambda < 0$, and we have linear combinations of cosine and sine, as follows:

$$\phi(y) = c_1 \cos\left(\sqrt{\lambda}y\right) + c_2 \sin\left(\sqrt{\lambda}y\right). \tag{4.29}$$

In order to satisfy the boundary conditions, we evaluate $\phi(0)$ and $\phi(H)$:

$$\begin{aligned} &\phi(0) = c_1 = 0 \implies c_1 = 0 \\ &\phi(H) = c_2 \sin\left(\sqrt{\lambda}H\right) = 0 \implies \sqrt{\lambda}H = n\pi \implies \lambda = \left(\frac{n\pi}{H}\right)^2. \end{aligned} \tag{4.30}$$

The sine function is 0 for all integers of π. Therefore, $H\sqrt{\lambda} = n\pi \implies \lambda_n = \left(\frac{n\pi}{H}\right)^2$. The corresponding eigenfunctions are: $\phi_n(y) = \sin\left(\frac{n\pi y}{H}\right)$.

Case 2: λ < 0. In this case, $-\lambda > 0$. For ease of notation, let $\mu = \sqrt{-\lambda}$. Thus, we have linear combinations of exponentials, as follows:

$$\phi(y) = c_1 e^{\mu y} + c_2 e^{-\mu y}. \tag{4.31}$$

In order to satisfy the boundary conditions, we evaluate $\phi(0)$ and $\phi(H)$:

$$\begin{aligned} \phi(0) &= c_1 + c_2 = 0 \\ \phi(H) &= c_1 e^{\mu H} + c_2 e^{-\mu H} = 0. \end{aligned} \tag{4.32}$$

This corresponds to the following system of equations:

$$\begin{bmatrix} 1 & 1 \\ e^{\mu H} & e^{-\mu H} \end{bmatrix} \begin{bmatrix} c_1 \\ c_2 \end{bmatrix} = \begin{bmatrix} 0 \\ 0 \end{bmatrix}. \tag{4.33}$$

The determinant of the coefficient matrix is: $e^{-\mu H} - e^{\mu H}$. This determinant is only 0 when $e^{\mu H} = e^{-\mu H}$. But this only occurs when $\mu = 0$. However, since $\mu > 0$, we conclude that the determinant of the matrix in Equation (4.33) is never 0. Thus, the null space is empty, which means that the only solution to the matrix equation is $c_1 = c_2 = 0$. Hence, there is only the trivial solution when $\lambda < 0$.

Case 3: $\boldsymbol{\lambda = 0}$. In this case, the ODE becomes $\phi''(y) = 0$. Integrating twice leads to $\phi(y) = c_1 y + c_2$. In order to satisfy the boundary conditions, we evaluate $\phi(0)$ and $\phi(H)$:

$$\begin{aligned} \phi(0) &= c_2 = 0 \implies c_2 = 0 \\ \phi(H) &= c_1 H = 0 \implies c_1 = 0. \end{aligned} \tag{4.34}$$

Hence, there is only the trivial solution when $\lambda = 0$.

Thus, we have found that the eigenvalues for this problem satisfy $\lambda > 0$ and are given by $\lambda_n = \left(\frac{n\pi}{H}\right)^2$. Now, we will solve the x-dependent problem. The solutions to the x-dependent problem are exponentials. However, we will write the solution as a linear combination of cosh and sinh, as these functions are themselves linear combinations of exponential functions. Observe:

$$\begin{aligned} \cosh\left(\sqrt{\lambda}x\right) &= \frac{e^{\sqrt{\lambda}x} + e^{-\sqrt{\lambda}x}}{2} \\ \sinh\left(\sqrt{\lambda}x\right) &= \frac{e^{\sqrt{\lambda}x} - e^{-\sqrt{\lambda}x}}{2}. \end{aligned} \tag{4.35}$$

Furthermore, we will translate the functions by L in order to satisfy the boundary condition at $x = L$. Therefore, the x-dependent ODE is satisfied by the following:

$$h(x) = d_1 \cosh\left[\sqrt{\lambda}(x - L)\right] + d_2 \sinh\left[\sqrt{\lambda}(x - L)\right]. \tag{4.36}$$

Now, we apply the boundary condition at $x = L$.

$$h'(L) = d_2\sqrt{\lambda} = 0 \implies d_2 = 0. \tag{4.37}$$

Thus, we have

$$u_n(x,y) = h_n(x)\phi_n(y) = d_1 \cosh\left[\sqrt{\lambda_n}(x-L)\right]\sin\left(\frac{n\pi y}{H}\right). \tag{4.38}$$

Furthermore, because the PDE is linear, the solution obeys the principle of superposition. Therefore,

$$u(x,y) = \sum_{n=1}^{\infty} c_n \cosh\left[\frac{n\pi}{H}(x-L)\right]\sin\left(\frac{n\pi y}{H}\right). \tag{4.39}$$

The values of the c_n will be determined in order to satisfy the non-homogeneous boundary condition. Thus, we evaluate $\frac{\partial u}{\partial x} = 0$:

$$\left.\frac{\partial u}{\partial x}\right|_{x=0} = \sum_{n=1}^{\infty} \frac{-c_n n\pi \sin\left(\frac{n\pi y}{H}\right)\sinh\left(\frac{Ln\pi}{H}\right)}{H} = 4y. \tag{4.40}$$

Re-arranging this expression leads to:

$$\sum_{n=1}^{\infty} c_n \sin\left(\frac{n\pi y}{H}\right) = -\frac{4Hy}{n\pi\sinh\left(\frac{Ln\pi}{H}\right)}. \tag{4.41}$$

To find the coefficients, multiply we invoke the orthogonality of sine functions of different modes:

$$\begin{aligned}
\int_0^H \sum_{n=1}^{\infty} c_n \sin\left(\frac{n\pi y}{H}\right)\sin\left(\frac{m\pi y}{H}\right)dy &= -\frac{4H}{n\pi\sinh\left(\frac{Ln\pi}{H}\right)}\int_0^H y\sin\left(\frac{n\pi y}{H}\right)dy \\
\sum_{n=1}^{\infty}\int_0^H c_n \sin\left(\frac{n\pi y}{H}\right)\sin\left(\frac{m\pi y}{H}\right)dy &= -\frac{4H}{n\pi\sinh\left(\frac{Ln\pi}{H}\right)}\int_0^H y\sin\left(\frac{n\pi y}{H}\right)dy \\
c_n\frac{H}{2} &= -\frac{4H}{n\pi\sinh\left(\frac{Ln\pi}{H}\right)}\int_0^H y\sin\left(\frac{n\pi y}{H}\right)dy \\
c_n &= -\frac{8}{n\pi\sinh\left(\frac{Ln\pi}{H}\right)}\int_0^H y\sin\left(\frac{n\pi y}{H}\right)dy.
\end{aligned} \tag{4.42}$$

The integral on the right-hand side can be further evaluated to obtain:

$$
\begin{aligned}
c_n &= -\frac{8}{n\pi \sinh\left(\frac{Ln\pi}{H}\right)}\left[\frac{H^2(-n\pi\cos(n\pi) + \sin(n\pi)^{\to 0})}{n^2\pi^2}\right] \\
&= -\frac{8}{n\pi \sinh\left(\frac{Ln\pi}{H}\right)}\left[\frac{H^2(-1)^{n+1}}{n\pi}\right] \\
&= \frac{(-1)^n 8H^2}{n^2\pi^2 \sinh\left(\frac{Ln\pi}{H}\right)}.
\end{aligned}
\tag{4.43}
$$

Therefore, we conclude the following solution to this boundary value problem:

$$
u(x,y) = \sum_{n=1}^{\infty} c_n \cosh\left[\frac{n\pi}{H}(x-L)\right]\sin\left(\frac{n\pi y}{H}\right), \tag{4.44}
$$

where

$$
c_n = \frac{(-1)^n 8H^2}{n^2\pi^2 \sinh\left(\frac{Ln\pi}{H}\right)}. \tag{4.45}
$$

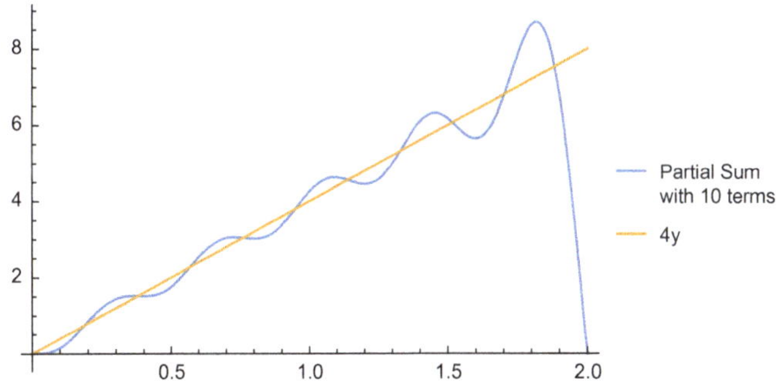

Figure 4.1: Evaluating $\left.\frac{\partial u}{\partial x}\right|_{x=0}$ using a partial sum of Equation (4.44) with 10 terms. For this example, we use $H = 2$.

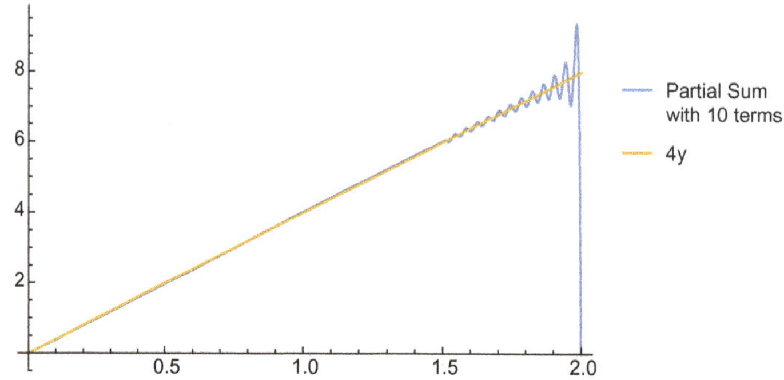

Figure 4.2: Evaluating $\left.\frac{\partial u}{\partial x}\right|_{x=0}$ using a partial sum of Equation (4.44) with 100 terms. For this example, we use $H = 2$.

Example 4.1.0.5

Consider the problem domain $\Omega = [0, L] \times [0, H]$. Suppose u satisfies Laplace's equation in Ω with boundary conditions given by:

$$\begin{aligned} u(0, y) = g(y) \quad &\text{and} \quad u(L, y) = 0 \\ u(x, 0) = 0 \quad &\text{and} \quad u(x, H) = 0. \end{aligned} \tag{4.46}$$

We seek to solve Laplace's equation inside a rectangle. In this case, the PDE is linear and homogeneous, however, the boundary condition imposed on the edge $x = 0$ is not homogeneous. For now, we will ignore the inhomogeneous boundary condition.

A Laplace BVP in a rectangle can be solved using a technique known as separation of variables. The strategy is to assume the solution is the product of a function solely of x and a function solely of y. Thus, we seek a product solution of the form

$$u(x, y) = h(x)\phi(y). \tag{4.47}$$

In terms of $h(x)$ and $\phi(y)$, the boundary conditions are $h(L) = 0$, $\phi(0) = 0$, and $\phi(H) = 0$. Substituting Equation (4.47) into Laplace's equation, we have

$$h_{xx}\phi + h\phi_{yy} = 0.$$

Separating this expression yields

$$\frac{1}{h}h_{xx} = -\frac{1}{\phi}\phi_{yy}.$$

The technique is known as separation of variables because the left-hand side is a function of x, and the right-hand side is a function of y. However, this implies that a function of x is equal everywhere to a function of y for all values of x and y. Thus, it must be the case that both the left and right sides of this equation are constant. This constant, $\lambda \in \mathbb{R}$, is known as the separation constant.

$$\frac{1}{h}h_{xx} = -\frac{1}{\phi}\phi_{yy} = \lambda.$$

This results in two ODEs, namely

$$h_{xx} = \lambda h$$

and

$$\phi_{yy} = -\lambda\phi.$$

Considering the boundary conditions, there are two homogeneous boundary conditions for the ϕ problem, but only one homogeneous boundary condition for the h problem. Thus, the x-dependent problem is not a boundary value problem since it does not have two homogeneous boundary conditions. So, we begin with the y-dependent problem. The process of checking each possible case for the eigenvalues was demonstrated in Example 4.1.0.4. In this case, this is a well-known problem and the eigenvalues and eigenfunctions are as follows:

$$\lambda_n = \left(\frac{n\pi}{H}\right)^2$$

and

$$\phi_n(y) = \sin\frac{n\pi y}{H}.$$

For more details on how the eigenvalues and eigenfunctions are obtained, the reader is referred to a standard text on partial differential equations, such as [16].

Now that we have an expression for the separation constants, λ_n, we can solve the x-dependent problem. We will assume the solution is an exponential function of the form $h(x) = e^{rx}$. Applying this assumed form to the ODE for h leads to the following characteristic equation:

$$r^2 e^{rx} = \lambda e^{rx} \implies r^2 - \lambda = 0.$$

Given the expression for the separation constants, it is clear that the separation constants are always positive. Thus, we have two real roots, namely $r = \pm\sqrt{\lambda}$. So, the general solution of the x-dependent problem can be expressed as either a linear combination of exponential functions or as a linear combination of hyperbolic functions (which are, themselves, linear combinations of exponential functions). In this case, in order to satisfy the condition $h(L) = 0$, we will use translated hyperbolic functions as follows:

$$h(x) = a_1 \cosh \frac{n\pi(x-L)}{H} + a_2 \sinh \frac{n\pi(x-L)}{H}.$$

We will quickly verify that this function satisfies the ODE for h:

$$\begin{aligned} h_{xx} &= \frac{a_1 n^2\pi^2}{H^2} \cosh \frac{n\pi(x-L)}{H} + \frac{a_2 n^2\pi^2}{H^2} \sinh \frac{n\pi(x-L)}{H} \\ &= \frac{n^2\pi^2}{H^2} h(x) \\ &= \lambda h. \end{aligned}$$

Now, upon applying the boundary condition $h(L) = 0$, we have

$$\begin{aligned} h(L) &= a_1 \cosh \frac{n\pi(L-L)}{H} + a_2 \sinh \frac{n\pi(L-L)}{H} \\ &= a_1 \cosh 0 + a_2 \sinh 0 \\ &= a_1 \implies a_1 = 0. \end{aligned}$$

Thus, the solution for the x-dependent problem is as follows:

$$h(x) = a_2 \sinh \frac{n\pi(x-L)}{H}.$$

Returning to the product solution $u(x,y) = h(x)\phi(y)$, we have:

$$u_n(x,y) = A_n \sin \frac{n\pi y}{H} \sinh \frac{n\pi(x-L)}{H}. \tag{4.48}$$

At this point, we can turn our attention to the inhomogeneous boundary condition. To satisfy this condition, we shall use the principle of superposition and the properties of the Fourier series to find coefficients such that the inhomogeneous boundary condition is satisfied. By the principle of superposition, since Equation (4.48) is a solution to the PDE, any linear combination of terms of the form in Equation (4.48) is also a solution. Thus,

$$u(x, y) = \sum_{n=1}^{\infty} A_n \sin \frac{n\pi y}{H} \sinh \frac{n\pi(x - L)}{H} \tag{4.49}$$

is also a solution of the PDE. We shall now use Equation (4.49) to satisfy the inhomogeneous boundary condition using the Fourier series. The inhomogeneous boundary condition corresponds to $x = 0$, so, evaluating at $x = 0$, we have

$$u(0, y) = \sum_{n=1}^{\infty} A_n \sin \frac{n\pi y}{H} \sinh \frac{n\pi(-L)}{H} = g(y). \tag{4.50}$$

Notice that this equation is just a function of y. So, if we consider the coefficient of the sine term to be $A_n \sinh \frac{-n\pi L}{H}$, we can use our knowledge of the orthogonality of sines of different modes to determine the coefficient values. The coefficients of the Fourier-sine series are given by

$$A_n \sinh \frac{n\pi(-L)}{H} = \frac{2}{H} \int_0^H g(y) \sin\left(\frac{n\pi y}{H}\right) dy.$$

Based on the properties of the hyperbolic sine function, $\sinh \frac{n\pi(-L)}{H}$ is never equal to zero. Thus, division by it is defined. So,

$$A_n = \frac{2}{H \sinh \frac{n\pi(-L)}{H}} \int_0^H g(y) \sin\left(\frac{n\pi y}{H}\right) dy.$$

Thus, the solution to this problem is

$$u(x, y) = \sum_{n=1}^{\infty} A_n \sin \frac{n\pi y}{H} \sinh \frac{n\pi(x - L)}{H}$$

where each A_n is given by

$$A_n = \frac{2}{H \sinh \frac{n\pi(-L)}{H}} \int_0^H g(y) \sin\left(\frac{n\pi y}{H}\right) dy.$$

Example 4.1.0.6

We now solve Laplace's equation inside a quarter-circle in the first quadrant of radius 1 ($0 \le \theta \le \frac{\pi}{2}, 0 \le r \le 1$). The geometry of this problem lends itself to the use of polar coordinates. Thus, for convenience, the Laplace equation in polar coordinates is

$$\nabla^2 u = \frac{1}{r}\frac{\partial}{\partial r}\left(r\frac{\partial u}{\partial r}\right) + \frac{1}{r^2}\frac{\partial^2 u}{\partial \theta^2} = 0. \tag{4.51}$$

In order to utilize the technique of separation of variables, we add some homogeneous subsidiary conditions that are motivated by the physical properties of the problem. Recall from when the Laplace equation was solved in the previous problem that boundary conditions are needed at the endpoints of the definitions of the variables. Thus, assuming that the process for solving the Laplace equation inside a circular disk is similar, we need boundary conditions at the endpoints of the definitions of the variables r and θ. Thus, we need boundary conditions at $r = 0$ and $r = 1$ as well as at $\theta = \pi/2$ and $\theta = -\pi/2$. The boundary condition at $r = 1$ is already known. Likewise, the condition at $\theta = \pi/2$ is already known. Because of the physical nature of the problem, we assume that the boundary condition at $\theta = \pi/2$ is equal to the boundary condition at $\theta = -\pi/2$. Thus, the only boundary condition that remains to be given is the condition at $r = 0$. Because polar coordinates are singular at this point, we specify that the temperature must be finite at the origin, thus, $|u(0,\theta)| < \infty$. As currently written, this problem is suited for separation of variables.

The boundary conditions of this problem correspond to the situation in which the parts of the quarter circle adjacent to the x and y axes (i.e. $\theta = 0$ and $\theta = \pi/2$, respectively) are insulated (Neumann conditions). The boundary conditions along the arc of the quarter circle are Dirichlet conditions and are specified by $f(\theta)$. We shall assume a solution that has a product form, and then we shall show that the problem is indeed separable.

$$u(r,\theta) = \phi(\theta)G(r).$$

Substituting this equation into the Laplace equation in polar coordinates, we have

$$\frac{1}{r}\frac{d}{dr}\left(rG'\right)\phi + \frac{1}{r^2}\phi''G = 0.$$

Separating this equation yields

$$\frac{r}{G}\frac{d}{dr}(rG') = -\frac{1}{\phi}\phi''.$$

However, this equation implies that r is a function of θ and that θ is a function of r. However, r and θ are both supposed to be independent variables that are able to assume any values. Thus, it must be the case that both the left and right sides of this equation must be equal to a constant. Thus,

$$\frac{r}{G}(rG'' + G') = -\frac{1}{\phi}\phi'' = \lambda.$$

This expression yields two ODEs, namely,

$$r^2G'' + rG' - \lambda G = 0$$

and

$$\phi'' + \lambda\phi = 0.$$

For the θ-dependent problem, the boundary conditions are given by

$$\phi'(0) = 0$$
$$\phi'\left(\frac{\pi}{2}\right) = 0$$

The θ-dependent problem has two homogeneous boundary conditions and is, consequently, an eigenvalue problem. The characteristic equation of the ODE in θ is

$$r^2 + \lambda = 0.$$

This implies that $r = \pm\sqrt{-\lambda}$. We need to consider the cases $\lambda = 0$, $\lambda > 0$, and $\lambda < 0$ separately. We begin with the case $\lambda = 0$. Thus, we have a general solution

$$\phi = c_1 + c_2\theta.$$

Differentiating yields $\phi' = c_2$. Applying the boundary conditions, it follows that $c_2 = 0$. Thus, $\phi = c_1$ is a general solution to the ODE. We now consider the case $\lambda > 0$. In this case, the roots of the characteristic equation are complex. Thus, the resulting general solution is a linear combination of cosines and sines, specifically,

$$\phi = c_1 \cos\sqrt{\lambda}\theta + c_2 \sin\sqrt{\lambda}\theta.$$

Differentiating yields $\phi'(\theta) = -c_1\sqrt{\lambda}\sin\sqrt{\lambda}\theta + c_2\sqrt{\lambda}\cos\sqrt{\lambda}\theta$. Applying the first boundary condition, we have

$$\phi'(0) = -c_1\sqrt{\lambda}(0) + c_2\sqrt{\lambda} = 0,$$

which implies that $c_2 = 0$ since $\lambda > 0$. Applying the second boundary condition, we have

$$\phi'\left(\frac{\pi}{2}\right) = -c_1\sqrt{\lambda}\sin\left(\sqrt{\lambda}\frac{\pi}{2}\right) = 0.$$

This has nontrivial solutions whenever $\sin\left(\sqrt{\lambda}\frac{\pi}{2}\right) = 0$. So, there are nontrivial solutions whenever

$$\sqrt{\lambda}\frac{\pi}{2} = n\pi.$$

Thus, the eigenvalues for this problem are

$$\lambda_n = (2n)^2, \text{ where } n = 1, 2, 3, \ldots.$$

The corresponding eigenfunctions are

$$\phi_n = c_1 \cos 2n\theta, \text{ where } n = 1, 2, 3, \ldots.$$

Lastly, we consider the case when $\lambda < 0$. In this case, the roots of the characteristic equation are real and distinct. Thus, the general solution is a linear combination of exponential functions. Namely,

$$\phi = c_1 e^{\sqrt{-\lambda}\theta} + c_2 e^{-\sqrt{-\lambda}\theta}.$$

Differentiating yields

$$\phi' = c_1\sqrt{-\lambda}e^{\sqrt{-\lambda}\theta} - c_2\sqrt{-\lambda}e^{-\sqrt{-\lambda}\theta}.$$

Applying the first boundary condition, we have $\phi'(0) = c_1\sqrt{-\lambda} - c_2\sqrt{-\lambda} = 0$, which implies that $c_1 = c_2$. Applying the second boundary condition, we have $\phi'(\pi/2) = c_1\sqrt{-\lambda}e^{\sqrt{-\lambda}\pi/2} - c_2\sqrt{-\lambda}e^{-\sqrt{-\lambda}\pi/2} = 0$. Since $\sqrt{-\lambda} \neq 0$, we have $c_1 e^{\sqrt{-\lambda}\pi/2} = c_2 e^{-\sqrt{-\lambda}\pi/2}$. Since $c_1 = c_2$, we have $c_1 e^{\sqrt{-\lambda}\pi/2} = c_1 e^{-\sqrt{-\lambda}\pi/2}$, which is only true if $c_1 = c_2 = 0$. Thus, this case only corresponds to trivial solutions.

As previously stated, the r-dependent problem is

$$r^2 G'' + rG' - \lambda G = 0.$$

This equation is linear and homogeneous, but it does not have constant coefficients. These types of problems are generally difficult to solve, however, this equation, which is called the Euler equation, does have a known solution. The general solution is given by

$$G = c_1 r^{\sqrt{\lambda}} + c_2 r^{-\sqrt{\lambda}}.$$

Based on the eigenvalues discovered from solving the θ-dependent problem, we have to consider two cases, $\lambda = 0$ and $\lambda_n = (2n)^2$. In the case where $\lambda = 0$, we have the general solution

$$G = \bar{c}_1 + \bar{c}_2 \ln r.$$

The only homogeneous boundary condition for the r-dependent problem is that the solution is bounded at the origin. Since $\ln r$ is not bounded as $r \to 0$, it follows that $c_2 = 0$. Thus,

$$G(r) = c_1 r^{\sqrt{\lambda}}, \text{ where } \lambda \geq 0.$$

So, when accounting for both cases of eigenvalues, the product solution $u(r,\theta) = \phi(\theta)G(r)$ is

$$u(r,\theta) = A_0 r^0 + \sum_{n=1}^{\infty} A_n r^{2n} \cos 2n\theta = \sum_{n=0}^{\infty} A_n r^{2n} \cos 2n\theta.$$

We now need to apply the inhomogeneous boundary condition. So,

$$u(1,\theta) = \sum_{n=0}^{\infty} A_n (1)^{2n} \cos 2n\theta = \sum_{n=0}^{\infty} A_n \cos 2n\theta = f(\theta).$$

To determine the coefficients A_n, we write this equation in the form of a Fourier-cosine series. Thus,

$$\sum_{n=0}^{\infty} A_n \cos 2n\theta = \sum_{n=0}^{\infty} A_n \cos \frac{n\pi\theta}{\frac{\pi}{2}} = f(\theta).$$

The coefficients for this Fourier-cosine series are given by

$$A_n = \frac{2}{\pi/2} \int_0^{\frac{\pi}{2}} f(\theta) \cos \frac{n\pi\theta}{\pi/2} d\theta.$$

Simplification yields

$$A_n = \frac{4}{\pi} \int_0^{\frac{\pi}{2}} f(\theta) \cos(2n\theta) d\theta.$$

Thus, the solution to the problem is

$$u(r,\theta) = \sum_{n=0}^{\infty} A_n r^{2n} \cos 2n\theta$$

where the A_n are given by

$$A_n = \frac{4}{\pi} \int_0^{\frac{\pi}{2}} f(\theta) \cos(2n\theta) d\theta.$$

The important observation to take from this section is as follows. Let Ω represent a problem domain with boundary $\partial\Omega$. If a function, u, satisfies the given PDE in Ω and satisfies the boundary conditions on $\partial\Omega$, then u is a solution to that boundary value problem.

Note 4.1.0.1

The Laplace operator is also a fundamental component of other common PDEs such as the heat equation and the wave equation. These PDEs are as follows:

$$\begin{aligned} \text{Heat equation: } & \frac{\partial u}{\partial t} = \frac{\partial^2 u}{\partial x^2} + \frac{\partial^2 u}{\partial y^2} = \Delta u \\ \text{Wave equation: } & \frac{\partial^2 u}{\partial t^2} = \frac{\partial^2 u}{\partial x^2} + \frac{\partial^2 u}{\partial y^2} = \Delta u. \end{aligned} \tag{4.52}$$

Solutions to Laplace's equation can be used in constructing solutions to the heat and wave equations. In particular, once a particular solution is found that satisfies the time component of the heat and wave equation problems, solutions to Laplace's equation can be used to satisfy the boundary conditions of these problems.

Furthermore, the Laplace equation is sometimes called the steady-state heat equation because Laplace's equation is equal to the heat equation when $\frac{\partial u}{\partial t} = 0$. When a time derivative such as this is equal to 0, the phenomenon being modeled is said to be at steady state because it is no longer changing with respect to time.

The papers [17], [18], and [19] have examined solving the heat and wave equations by finding a particular solution for the temporal component and using the CVBEM to solve the boundary value problem.

4.2 Introduction to the CVBEM

In this section:

1. Statement of the CVBEM approximation function.
2. Discussion of degrees of freedom.
3. Special properties of the CVBEM approximation function that are a consequence of using analytic complex variable basis functions.

The CVBEM is a numerical method for solving boundary value problems of the two-dimensional Laplace type. The CVBEM utilizes the properties of analytic complex variable functions to obtain an approximation function with real and imaginary parts that are harmonic within a specified problem domain.

Problem Statement Consider a region $\Omega_R \subset \mathbb{R}^2$ in which Laplace's equation holds. We assume that Ω_R is a simply connected, open subset of

$\mathbb{R}^2$. The boundary of Ω_R is denoted $\Gamma_R = \partial\Omega_R$ and is positively oriented. Boundary conditions are specified on Γ_R.

Definition 4.2.0.1 We define the **CVBEM approximation function**, denoted $\hat{\omega}(z)$, as a linear combination of analytic complex variable functions with the following form:

$$\hat{\omega}(z) = \sum_{j=1}^{n} c_j g_j(z), \tag{4.53}$$

where $c_j \in \mathbb{C}$ denotes a complex coefficient, $g_j(z)$ denotes a complex variable function that is analytic at least within the problem domain, and n is the number of basis functions being used in the approximation.

The complex coefficients used in Equation (4.53) are composed of two real constants:

$$c_j = \alpha_j + i\beta_j, \quad \mathrm{Re}(c_j) = \alpha_j, \quad \mathrm{Im}(c_j) = \beta_j. \tag{4.54}$$

The real and the imaginary components of each c_j are degrees of freedom that need to be determined as part of the CVBEM modeling process. Thus, since there are n complex coefficients in the linear combination of the CVBEM approximation function, it follows that there are $2n$ degrees of freedom to be determined in a CVBEM approximation function.

Note 4.2.0.1

Various techniques for determining the coefficients of the CVBEM approximation function are examined in the literature. These techniques include collocation and least squares using the given boundary data, among others [15].

Note 4.2.0.2

Since the CVBEM approximation function, $\hat{\omega}$, is a linear combination of complex variable functions that are each analytic within the problem domain, $\hat{\omega}$ is also analytic within the problem domain by Theorem 2.3.4.2.

Therefore, both the real and imaginary parts of the CVBEM approximation function are harmonic functions. As a result, the CVBEM approximation function satisfies the governing Laplace PDE within the problem domain without any error.

So, the primary goal of modeling with the CVBEM is to determine the α_j and β_j such that the resulting approximation function agrees with the specified boundary conditions as closely as possible.

As indicated in Equation (4.54), each c_j is composed of a real and an imaginary part, α_j and β_j, respectfully. Thus, Equation (4.53) may be written equivalently as:

$$\hat{\omega}(z) = \sum_{j=1}^{n} \left(\alpha_j + i\beta_j\right) g_j(z).$$

Likewise, the basis functions, $g_j(z)$, are composed of real and imaginary parts: $g_j(z) = \lambda_j(z) + i\mu_j(z)$. Thus,

$$\hat{\omega}(z) = \sum_{j=1}^{n} \left(\alpha_j + i\beta_j\right)\left[\lambda_j(z) + i\mu_j(z)\right].$$

After multiplying these terms, we obtain:

$$\hat{\omega}(z) = \sum_{j=1}^{n} \left[\alpha_j\lambda_j(z) - \beta_j\mu_j(z) + i\alpha_j\mu_j(z) + i\beta_j\lambda_j(z)\right].$$

Thus, the real part of the CVBEM approximation function is

$$\mathrm{Re}\left[\hat{\omega}(z)\right] = \hat{\phi}(z) = \sum_{j=1}^{n} \left[\alpha_j\lambda_j(z) - \beta_j\mu_j(z)\right]. \tag{4.55}$$

Likewise, the imaginary part of the CVBEM approximation function is

$$\mathrm{Im}\left[\hat{\omega}(z)\right] = \hat{\psi}(z) = \sum_{j=1}^{n} \left[\alpha_j\mu_j(z) + \beta_j\lambda_j(z)\right]. \tag{4.56}$$

Note 4.2.0.3

As indicated in Equations (4.55) and (4.56), both the real and imaginary parts of the CVBEM approximation function share the same coefficients; namely, the α_j and β_j. These coefficients are determined so as to minimize the error of the CVBEM approximation function in satisfying the given boundary conditions.

Since $\hat{\phi}$ and $\hat{\psi}$ share the same coefficients, as soon as one of those two functions is determined, the other is immediately known. This is a consequence of the fact that the real and imaginary parts of an analytic complex variable function are related by the Cauchy–Riemann equations. Thus, once either the real or imaginary part is known, the other can be determined uniquely up to an additive constant.

The CVBEM approach yields an approximation function that is analytic throughout the problem domain. Therefore, the real and imaginary parts of the approximation function are continuously defined within the problem domain and, moreover, are harmonic functions within the problem domain. These properties constitute a considerable advantage over the common domain discretization methods such as the finite element method and finite difference method. Such domain discretization methods only provide approximations of the solution at a finite number of mesh points within the problem domain and do not typically yield an approximation function satisfying the strong form of the governing PDE.

Note 4.2.0.4

The analytic basis functions that are used in the CVBEM approximation function can be selected by the user and their selection may be motivated by such considerations as the geometry of the problem boundary, the nature of the specified boundary conditions, or the computational convenience of the basis functions, among other factors. We examine different basis functions that have successfully been used in CVBEM formulations in Section 4.3.

4.3 CVBEM basis functions

In this section:

1. We will derive basis functions for the CVBEM based on the Cauchy integral equation for analytic complex variable functions. Straight-line segments are used to discretize the problem boundary.
2. We will show that the Cauchy integral equation can be directly integrated along these segments. The ability to perform direct integration along the boundary elements eliminates the need to use quadrature formulae, which is a necessary feature of many comparable real variable boundary element methods.
3. We will also present other analytic basis functions that the authors have successfully used in various implementations of the CVBEM.

Note 4.3.0.1

Numerical methods requiring discretization of the problem domain, such as the finite element method or finite difference method, to model a two-dimensional BVP, require the user to create a mesh of the problem domain using two-dimensional elements. However, by employing a boundary element method, such as the CVBEM, it is only necessary to discretize the boundary, which results in a one-dimensional reduction of the meshing requirements for the problem at hand. For this reason, CVBEM programs typically require less time and storage resources dedicated to the meshing process compared with similar domain discretization methods solving the same problem.

For convenience, we repeat the problem statement given in Section 4.2.

Problem Statement Consider a region $\Omega_R \subset \mathbb{R}^2$ in which Laplace's equation holds. We assume that Ω_R is a simply connected, open subset of $\mathbb{R}^2$. The boundary of Ω_R is denoted $\Gamma_R = \partial\Omega_R$ and is positively oriented. Boundary conditions are specified on Γ_R.

In order to model this problem using the CVBEM, the problem domain and problem boundary are mapped into the complex plane. The images in $\mathbb{C}$ are denoted $\Omega \subset \mathbb{C}$ and $\Gamma \subset \mathbb{C}$, respectively. The sets Ω and Γ are related to Ω_R

and Γ_R by the map $f : \Omega_R \cup \Gamma_R \to \Omega \cup \Gamma$ that is given by $f(x, y) = x + iy$. The sets Ω and Γ are related by $\Gamma = \partial\Omega$.

With this information, we begin our development of the analytic CVBEM basis functions that are obtained by directly integrating the Cauchy integral equation. The following development is adapted from [3] and [20].

Let $\omega(z) = \phi(x, y) + i\psi(x, y)$ be analytic in $\Omega \cup \Gamma$. As demonstrated in Section 2.3.3, the real and imaginary components of ω are related by the Cauchy–Riemann equations:

$$\frac{\partial\phi}{\partial x} = \frac{\partial\psi}{\partial y} \quad \text{and} \quad \frac{\partial\phi}{\partial y} = -\frac{\partial\psi}{\partial x}.$$

Consequently, as shown in Section 2.3.4, ϕ and ψ are harmonic and satisfy Laplace's equation within Ω. Thus,

$$\frac{\partial^2\xi}{\partial x^2} + \frac{\partial^2\xi}{\partial y^2} = 0, \quad z \in \Omega \cup \Gamma, \quad \xi = \phi, \psi$$

Note 4.3.0.2

Below is a list of the major symbology used in this section:

Symbol	Meaning
Ω	Problem domain
$\hat{\Omega}$	Polygonal approximation of problem domain
$\Gamma = \partial\Omega$	Boundary of problem domain
$\hat{\Gamma} = \partial\hat{\Omega}$	Boundary of polygonal approximation of problem domain
m	The number of boundary elements
$\hat{\Gamma}_j, j = 1, \ldots, m$	The boundary elements
z	Complex variable

Symbol	Meaning
$\omega(z)$	An analytic complex variable function defined in Ω
$\hat{\omega}(z)$	CVBEM approximation of ω defined in $\hat{\Omega}$
$z_j, j = 1, \ldots, m$	Nodal end-points
$(k + 1)$	The number of nodal points defined on each $\hat{\Gamma}_j$
$\varsigma(s)$	Local coordinate system with parameter, s
$N^k_{j,i}(s)$, $j = 1, \ldots, m$ $i = 1, \ldots, (k + 1)$	Polynomial shape elements function of degree k

Note 4.3.0.3

By convention, ϕ is known as the potential (or state) function, and ψ is known as the stream function.

Due to the relationship described by the Cauchy–Riemann equations, ϕ and ψ are a special pair of harmonic functions known as conjugate functions. Since ϕ and ψ are conjugate functions, their contours, which are lines of equipotential and streamlines, respectively, are orthogonal at their points of intersection. This property provides for the simple creation of a flow net, which is an important visualization tool in the analysis of many physical phenomena governed by Laplace's equation.

Example 4.3.0.1

In this example, we demonstrate one of the important properties of conjugate harmonic functions. Namely, that the contour lines of conjugate harmonic functions are orthogonal at their points of intersection. This property is used to create flow nets, which can be useful for visualizing two-dimensional potential flow. Two examples of flow nets are presented below.

In the first example, we consider the complex variable function z^2, which is analytic on $\mathbb{C}$. The conjugate harmonic real variable functions corresponding to this complex variable function are as follows:

$$\begin{aligned}\phi(x,y) &= \mathrm{Re}\left(z^2\right) = \mathrm{Re}\left[(x+iy)^2\right] = \mathrm{Re}\left[x^2+2ixy-y^2\right]\\ &= x^2-y^2\\ \psi(x,y) &= \mathrm{Im}\left(z^2\right) = \mathrm{Im}\left[(x+iy)^2\right] = \mathrm{Im}\left[x^2+2ixy-y^2\right].\\ &= 2xy\end{aligned}$$

In the second example, we consider the complex variable function $z+\frac{1}{z}$, which is analytic on the set $\mathbb{C}\setminus\{0\}$. The conjugate harmonic real variable functions corresponding to this complex variable function are as follows:

$$\phi(x,y) = \text{Re}\left(z + \frac{1}{z}\right) = \text{Re}\left(x + iy + \frac{1}{x+iy}\right) = \text{Re}\left(x + iy + \frac{x-iy}{x^2+y^2}\right)$$
$$= x + \frac{x}{x^2+y^2}$$
$$\psi(x,y) = \text{Im}\left(z + \frac{1}{z}\right) = \text{Im}\left(x + iy + \frac{1}{x+iy}\right) = \text{Im}\left(x + iy + \frac{x-iy}{x^2+y^2}\right)$$
$$= y - \frac{y}{x^2+y^2}.$$

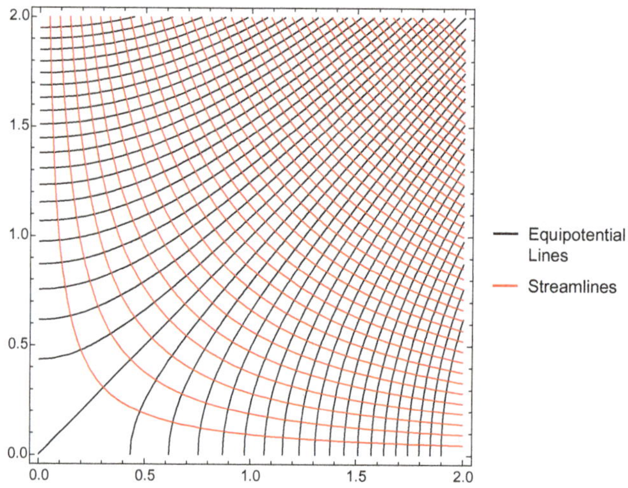

Figure 4.3: Example of equipotential lines and streamlines. Notice that the lines of equipotential and streamlines are orthogonal at their points of intersection. In this example, the potential function is $\phi(x,y) = \text{Re}\left(z^2\right) = x^2 - y^2$, and the stream function is $\psi(x,y) = \text{Im}\left(z^2\right) = 2xy$. The complex variable function z^2 is used to model ideal fluid flow in a 90-degree bend. This particular flow situation will be modeled later in Chapter 5.

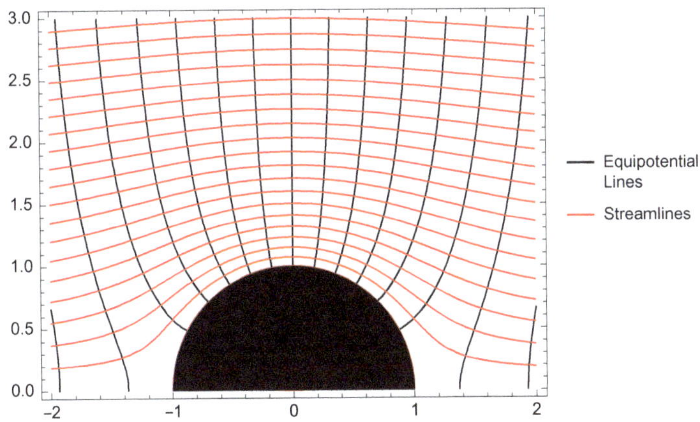

Figure 4.4: Example of equipotential lines and streamlines. Notice that the lines of equipotential and streamlines are orthogonal at their points of intersection. In this example, the potential function is $\phi(x,y) = \text{Re}\left(z + \frac{1}{z}\right) = x + \frac{x}{x^2+y^2}$, and the stream function is $\psi(x,y) = \text{Im}\left(z + \frac{1}{z}\right) = y - \frac{y}{x^2+y^2}$. The complex variable function z^2 is used to model ideal fluid flow in a 90-degree bend. This particular flow situation will be modeled later in Chapter 5.

4.3.1 Boundary elements

Now, recall the Cauchy integral formula for analytic complex variable functions:

$$\omega(z) = \frac{1}{2\pi i} \int_{\Gamma} \frac{\omega(\zeta)\,d\zeta}{\zeta - z}, \quad z \in \Omega. \tag{4.57}$$

Note 4.3.1.1

If a complex variable function, ω, is analytic within $\Omega \cup \Gamma$, and if the values of ω are known continuously along Γ, then the Cauchy integral formula can be used to determine $\omega(z_0)$ for all $z_0 \in \Omega$. Thus, every value of ω can be determined within Ω using only knowledge of the values of ω on Γ and the fact that ω is analytic within Ω.

To begin modeling with the CVBEM, it is necessary to approximate Γ. The approximation of Γ, denoted $\hat{\Gamma}$, is created by defining a set of m nodal points $\{z_1, \ldots, z_m\}$ on Γ. The points are labeled by first identifying z_1 (this point may be arbitrarily selected on Γ), and then proceeding around the boundary in the positive sense while labeling the subsequent nodal points with increasing indices, as depicted in Figure 4.5.

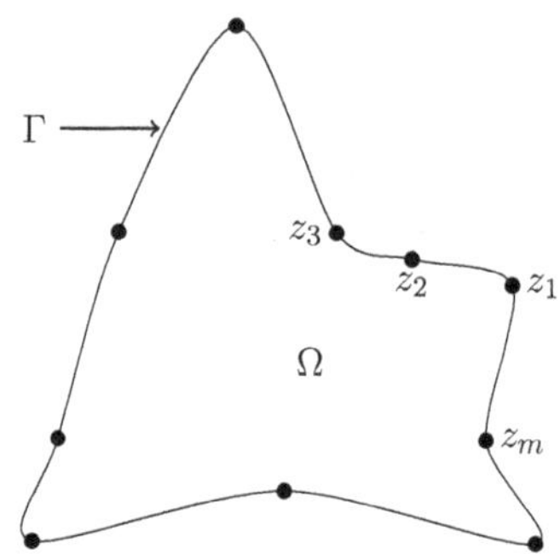

Figure 4.5: To begin modeling, m nodal points are defined on the problem boundary. The particular nodal point that is labeled as z_1 can be arbitrarily selected, but the rest of the nodal points are labeled in order starting at z_1 and proceeding in the counter-clockwise (positive) direction around the problem boundary until arriving at the final nodal point, which is denoted as z_m.

Definition 4.3.1.1 The boundary nodal points, $z_1, z_2, z_3, \ldots, z_m$, that are depicted in Figure 4.5 are known as **element end-point nodes.**

To create $\hat{\Gamma}$, each element end-point node is connected to exactly one other element end-point node with a linear curve so that there are m line segments, denoted $\hat{\Gamma}_1, \ldots, \hat{\Gamma}_m$.

Definition 4.3.1.2 The collection of $\hat{\Gamma}_j$ are known as **boundary elements.** The boundary elements are defined as

$$\hat{\Gamma}_j = \{z : z(s) = z_j(1-s) + z_{j+1}(s), \quad 0 \le s < 1\},$$
$$j = 1, \ldots, m-1 \tag{4.58}$$
$$\hat{\Gamma}_m = \{z : z(s) = z_m(1-s) + z_1(s), \quad 0 \le s < 1\}$$

where the z_j are the element end-point nodes defined on Γ.

By Definition 4.3.1.2, each $\hat{\Gamma}_j$ is the line segment whose endpoints are z_j and z_{j+1}, where it is understood that $z_{m+1} \equiv z_1$.

Note 4.3.1.2

By definition of the boundary elements in Definition 4.3.1.2, the following is true:

$$z_j \in \hat{\Gamma}_j \quad \text{and} \quad z_{j+1} \notin \hat{\Gamma}_j.$$

Therefore, $\Gamma_j \cap \Gamma_i = \varnothing$ for all $j \neq i$. So, the approximate boundary is defined as $\hat{\Gamma} = \bigcup_{j=1}^{m} \hat{\Gamma}_j$. Consequently, the set $\left\{\hat{\Gamma}_1, \ldots, \hat{\Gamma}_m\right\}$ is a partition of $\hat{\Gamma}$.

In Figure 4.6, we demonstrate how the specification of element end-point nodes on Γ is used to create the boundary elements $\hat{\Gamma}_j$.

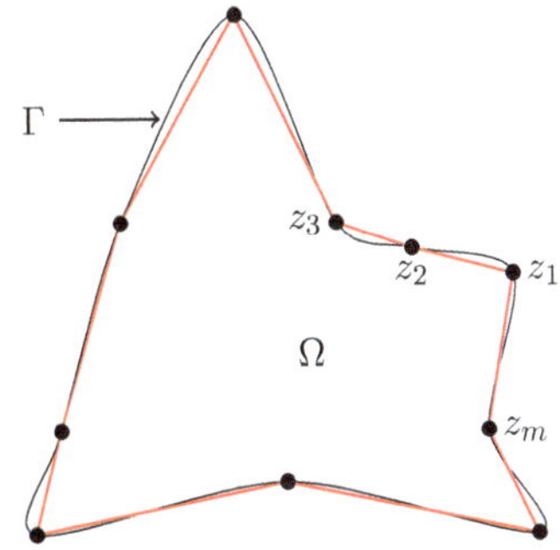

Figure 4.6: The problem boundary is depicted by the black curve. The approximate problem boundary is depicted by the red linear segments. The endpoints of each segment correspond to boundary nodal points, which are depicted as black points.

Of course, a better approximation of the problem boundary can be obtained by utilizing more element end-point nodes—and, consequently, more boundary elements—particularly in areas where the problem boundary is highly non-linear, or in other areas where there are large discrepancies, in a geometric sense, between $\hat{\Gamma}$ and Γ.

Definition 4.3.1.3 When discretizing the problem boundary, it is possible to define additional boundary nodal points in the interior of any of the boundary elements. When these types of nodes are utilized, we distinguish them from the element end-point nodes of Definition 4.3.1.1 by referring to them as **element interior nodes** or **interior nodal points**.

Element interior nodes are depicted graphically as circles that are not filled in, as illustrated in Figure 4.7.

In a CVBEM model, each boundary element can contain none, one, or several interior nodal points. Let each $\hat{\Gamma}_j$ be discretized by $(k+1)$ nodes (with $k \geq 1$). The quantity $(k+1)$ includes both element nodes as well as element interior nodes. Then, $\hat{\Gamma}_j$ is said to be a $(k+1)$-node element. Further, we may define for each $\hat{\Gamma}_j$ an associated nodal coordinate system such that $z_{j,1} = z_j$ and $z_{j,k+1} = z_{j+1} = z_{j+1,1}$.

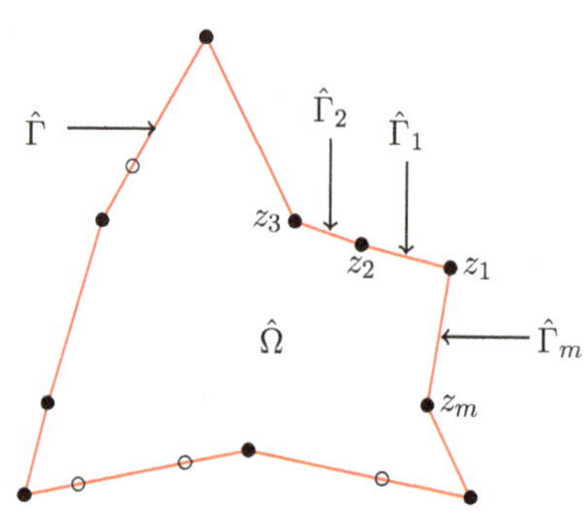

Figure 4.7: The polygonal approximation of the target problem boundary. The labeling scheme for the boundary nodal points is shown as well as the labeling scheme for the line segments $\hat{\Gamma}_1, \ldots, \hat{\Gamma}_2$. Also, note that it is possible to define additional boundary nodal points in the interior of any of the $\hat{\Gamma}_j$. These points are called element interior nodes, and are depicted by circles that are not filled in.

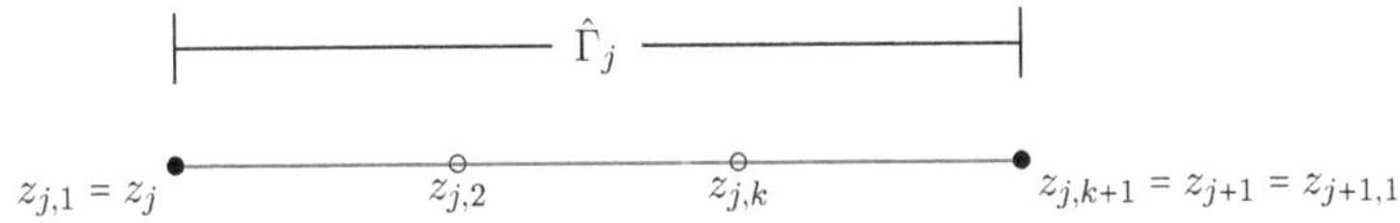

Figure 4.8: A boundary element depicting the use of element end-point nodes as well as element interior nodes. The nodal coordinate system provides a way for labeling and referring to element interior nodes.

Note 4.3.1.3

The significance of the nodal points, including both element end-point nodes and element interior nodes, is that information about $\omega(z) = \phi(z) + i\psi(z)$ is known at the nodal points. Let $z_{j,i}$ denote an arbitrary nodal point. At each $z_{j,i}$, either $\phi(z_{j,i})$ or $\psi(z_{j,i})$, or both, will be known.

With this discretization of the problem boundary in place, we return to the Cauchy integral equation, which was given in Equation (4.57). The exact integral expression for $\omega(z)$ is now replaced with the following approximation, denoted $\bar{\omega}(z) \approx \omega(z)$, which is based on the polygonal approximate boundary, $\hat{\Gamma}$.

$$\omega(z) = \frac{1}{2\pi i}\int_{\Gamma} \frac{\omega(\zeta)\,\mathrm{d}\zeta}{\zeta - z}, \; z \in \Omega \quad \longrightarrow \quad \bar{\omega}(z) = \frac{1}{2\pi i}\int_{\hat{\Gamma}} \frac{\omega(\zeta)\,\mathrm{d}\zeta}{\zeta - z}, \; z \in \hat{\Omega}$$

Next, we can express the integral over $\hat{\Gamma}$ as a sum of integrals over each of the boundary elements, $\hat{\Gamma}_j$, as follows:

$$\bar{\omega} = \frac{1}{2\pi i}\int_{\hat{\Gamma}} \frac{\omega(\zeta)\,\mathrm{d}\zeta}{\zeta - z} = \frac{1}{2\pi i}\sum_{j=1}^{m}\left[\int_{\hat{\Gamma}_j} \frac{\omega(\zeta)\,\mathrm{d}\zeta}{\zeta - z}\right], \qquad z \in \hat{\Omega}.$$

Now, we introduce a linear local coordinate system:

$$\zeta(s) = z_j + (z_{j+1} - z_j)s, \quad 0 \le s \le 1 \tag{4.59}$$

Using this change of variables, it follows that $\mathrm{d}\zeta = (z_{j+1} - z_j)\,\mathrm{d}s$, and $\omega(\zeta) = \omega(\zeta(s))$. We denote $\omega(\zeta(s))$ by the more notationally convenient $[\omega(s)]_j$, where the subscript notation, $[\;]_j$, is used to indicate that the values of ω are only being considered on $\hat{\Gamma}_j$.

Thus, in the local coordinate system, we have

$$\begin{aligned}\bar{\omega}(z) &= \frac{1}{2\pi i}\sum_{j=1}^{m}\int_{0}^{1}\frac{\left[\omega(s)\right]_j (z_{j+1}-z_j)\,\mathrm{d}s}{\zeta(s)-z}, && z\in\hat{\Omega}\\ &= \frac{1}{2\pi i}\sum_{j=1}^{m}\int_{0}^{1}\frac{\left[\omega(s)\right]_j (z_{j+1}-z_j)\cdot\frac{1}{(z_{j+1}-z_j)}\,\mathrm{d}s}{\left(\zeta(s)-z\right)\cdot\frac{1}{(z_{j+1}-z_j)}}, && z\in\hat{\Omega}\\ &= \frac{1}{2\pi i}\sum_{j=1}^{m}\int_{0}^{1}\frac{\left[\omega(s)\right]_j\,\mathrm{d}s}{\frac{\zeta(s)-z}{(z_{j+1}-z_j)}}, && z\in\hat{\Omega}.\end{aligned}$$

Now, using the definition of $\zeta(s)$ given above in Equation (4.59), we obtain the following after algebraic simplifications:

$$\begin{aligned}\bar{\omega}(z) &= \frac{1}{2\pi i}\sum_{j=1}^{m}\int_{0}^{1}\frac{\left[\omega(s)\right]_j\,\mathrm{d}s}{\frac{(z_{j+1}-z_j)s+z_j-z}{(z_{j+1}-z_j)}}, && z\in\hat{\Omega}\\ &= \frac{1}{2\pi i}\sum_{j=1}^{m}\int_{0}^{1}\frac{\left[\omega(s)\right]_j\,\mathrm{d}s}{s+\frac{z_j-z}{z_{j+1}-z_j}}, && z\in\hat{\Omega}\\ &= \frac{1}{2\pi i}\sum_{j=1}^{m}\int_{0}^{1}\frac{\left[\omega(s)\right]_j\,\mathrm{d}s}{s+\chi_j(z)}, && z\in\hat{\Omega},\end{aligned}\tag{4.60}$$

where $\chi_j(z)=\frac{z_j-z}{z_{j+1}-z_j}$. Notice that $\chi_j(z)$ is not a function of the integration variable, s.

Now, recall that $\left[\omega(s)\right]_j$ is composed of real and imaginary parts such that:

$$\left[\omega(s)\right]_j=\left[\phi(s)\right]_j+i\left[\psi(s)\right]_j,\quad 0\le s\le 1,$$

where, as before, we have used the more notationally convenient $\left[\phi(s)\right]_j$ and $\left[\psi(s)\right]_j$ to represent the values of $\phi\left(\zeta(s)\right)$ and $\psi\left(\zeta(s)\right)$, respectively, on the boundary element $\hat{\Gamma}_j$.

Therefore, Equation (4.60) can be expressed as the following sum of two complex variable definite integrals:

$$\begin{aligned}\bar{\omega}(z) &= \frac{1}{2\pi i}\sum_{j=1}^{m}\int_{0}^{1}\frac{\left[\phi(s)\right]_j+i\left[\psi(s)\right]_j\,\mathrm{d}s}{s+\chi_j(z)}, && z\in\hat{\Omega}\\ &= \frac{1}{2\pi i}\sum_{j=1}^{m}\left\{\int_{0}^{1}\frac{\left[\phi(s)\right]_j\,\mathrm{d}s}{s+\chi_j(z)}+i\int_{0}^{1}\frac{\left[\psi(s)\right]_j\,\mathrm{d}s}{s+\chi_j(z)}\right\}, && z\in\hat{\Omega}.\end{aligned}\tag{4.61}$$

4.3.2 Shape functions

The values of $\omega(s)$ are only known at the nodal points. However, in order to use the Cauchy integral theorem, we must know the values of ω continuously along the problem boundary. Therefore, since ω is not actually known continuously along the problem boundary, it is necessary to approximate its values in between the nodal points, which are the only places where ω is actually known.

On each boundary element, $\hat{\Gamma}_j$, we will define shape functions, denoted $N_{j,i}^k(s)$. The shape functions are used to approximate ω using the known values of ω at the nodal points. The resulting approximation is denoted $\hat{\omega}$ and is defined as follows:

$$\left[\hat{\omega}(s)\right]_j \equiv \sum_{i=1}^{k+1} N_{j,i}^k(s)\omega_{j,i}, \qquad 0 \le s \le 1 \tag{4.62}$$

where $\omega_{j,i} = \omega(z_{j,i})$ denotes the value of ω at the ith nodal point of $\hat{\Gamma}_j$.

We define the CVBEM approximation function as a function of the shape functions, as follows:

$$\begin{aligned}\hat{\omega}(z) &\equiv \frac{1}{2\pi i}\sum_{j=1}^{m}\int_0^1 \frac{\left[\hat{\omega}(s)\right]_j \mathrm{d}s}{s+\chi_j(z)}, && z \in \hat{\Omega}\\ &= \frac{1}{2\pi i}\sum_{j=1}^{m}\int_0^1 \frac{\sum_{i=1}^{k+1} N_{j,i}^k(s)\omega_{j,i}\,\mathrm{d}s}{s+\chi_j(z)}, && z \in \hat{\Omega}\\ &= \frac{1}{2\pi i}\sum_{j=1}^{m}\sum_{i=1}^{k+1}\int_0^1 \frac{N_{j,i}^k(s)\omega_{j,i}\,\mathrm{d}s}{s+\chi_j(z)}, && z \in \hat{\Omega}\end{aligned} \tag{4.63}$$

Now that the CVBEM approximation function has been defined, the remainder of this section is dedicated to directly integrating the approximation function given in Equation (4.63). The purpose of directly integrating this function is to provide a more easily usable form of the CVBEM approximation function that does not require the use of numerical quadrature.

Recall that $\omega_{j,i} = \phi_{j,i} + i\psi_{j,i}$. Thus,

$$\begin{aligned}
\left[\hat{\omega}(s)\right]_j &\equiv \sum_{i=1}^{k+1} N_{j,i}^k(s)\omega_{j,i}, && 0 \le s \le 1 \\
&= \sum_{i=1}^{k+1} N_{j,i}^k(s)\left(\phi_{j,i} + i\psi_{j,i}\right), && 0 \le s \le 1 \\
&= \sum_{i=1}^{k+1} N_{j,i}^k(s)\phi_{j,i} + i\sum_{i=1}^{k+1} N_{j,i}^k(s)\psi_{j,i}, && 0 \le s \le 1.
\end{aligned}$$

Hence, we define the following approximations of $\phi(s)$ and $\psi(s)$:

$$\begin{aligned}
\left[\phi(s)\right]_j \approx \left[\hat{\phi}(s)\right]_j &= \mathrm{Re}\left\{\left[\hat{\omega}(s)\right]_j\right\} = \sum_{i=1}^{k+1} N_{j,i}^k(s)\phi_{j,i}, && 0 \le s \le 1 \\
\left[\psi(s)\right]_j \approx \left[\hat{\psi}(s)\right]_j &= \mathrm{Im}\left\{\left[\hat{\omega}(s)\right]_j\right\} = \sum_{i=1}^{k+1} N_{j,i}^k(s)\psi_{j,i}, && 0 \le s \le 1.
\end{aligned} \tag{4.64}$$

The functions $N_{j,i}^k(s)$ appearing in Equations (4.62) and (4.64) are polynomial functions of degree k and are known as shape functions. Additionally, since $\left[\hat{\phi}(s)\right]_j$ and $\left[\hat{\psi}(s)\right]_j$ are sums of polynomial functions of degree k, $\left[\hat{\phi}(s)\right]_j$ and $\left[\hat{\psi}(s)\right]_j$ are themselves polynomial functions of degree k. Thus, there exist constants $a_{j,0}, a_{j,1}, \ldots, a_{j,k}$ and $b_{j,0}, b_{j,1}, \ldots, b_{j,k}$ for which we may write the following equivalent form of Equation (4.64):

$$\begin{aligned}
\left[\hat{\phi}(s)\right]_j &= a_{j,0} + a_{j,1}s + \cdots + a_{j,k}s^k = \sum_{p=0}^{k} a_{j,p}s^p, && 0 \le s \le 1 \\
\left[\hat{\psi}(s)\right]_j &= b_{j,0} + b_{j,1}s + \cdots + b_{j,k}s^k = \sum_{p=0}^{k} b_{j,p}s^p, && 0 \le s \le 1.
\end{aligned} \tag{4.65}$$

Moreover, since $\left[\hat{\omega}(s)\right]_j = \left[\phi(s)\right]_j + i\left[\psi(s)\right]_j$, we can write

$$\left[\hat{\omega}(s)\right]_j = c_{j,0} + c_{j,1}s + \cdots + c_{j,k}s^k = \sum_{p=0}^{k} c_{j,p}s^p, \qquad 0 \le s \le 1, \tag{4.66}$$

where $c_{j,p} = a_{j,p} + ib_{j,p}$, $p = 0, 1, \ldots, k$.

Note 4.3.2.1

Pertaining to the CVBEM shape functions, the significant equations and their relationships are as follows:

$$\left[\hat{\omega}(s)\right]_j \equiv \sum_{i=1}^{k+1} N_{j,i}^k(s)\omega_{j,i} = \sum_{p=0}^{k} c_{j,p}s^p, \qquad 0 \le s \le 1$$

$$\left[\hat{\phi}(s)\right]_j = \text{Re}\left\{\left[\hat{\omega}(s)\right]_j\right\} = \sum_{p=0}^{k} a_{j,p}s^p, \qquad 0 \le s \le 1$$

$$\left[\hat{\psi}(s)\right]_j = \text{Im}\left\{\left[\hat{\omega}(s)\right]_j\right\} = \sum_{p=0}^{k} b_{j,p}s^p, \qquad 0 \le s \le 1$$

$$c_{j,p} = a_{j,p} + ib_{j,p}$$

Note 4.3.2.2

The shape functions referenced in Equations (4.62) and (4.64) are analogous to the one-dimensional interpolation functions used in finite element methods. Additionally, note that we have used different letters to denote the coefficients of $\left[\hat{\phi}(s)\right]_j$ and $\left[\hat{\psi}(s)\right]_j$ in Equation (4.65). This is because the values of the coefficients in this equation depend on the given values of $\phi_{j,i}$ and $\psi_{j,i}$.

The shape functions satisfy the following conditions:

1. The first condition is that each shape function attains a value of 1 at one, and only one, of the nodal points in a given boundary element, $\hat{\Gamma}_j$. The shape function is 0 at all of the other nodal points within the given boundary element. This condition is expressed as:

$$N_{j,i}^k\left(\frac{z_{j,n} - z_{j,1}}{z_{j,k+1} - z_{j,1}}\right) = \begin{cases} 1, & n = i \\ 0, & n \neq i \end{cases}$$

2. The second condition is that each shape function has local support such that it is only nonzero on one boundary element, denoted $\hat{\Gamma}_j$, and is identically zero on all other boundary elements. This condition is expressed as:

$$N_{j,i}^k\left(\frac{\xi - z_{j,1}}{z_{j,k+1} - z_{j,1}}\right) = \begin{cases} \frac{\xi - z_{j,1}}{z_{j,k+1} - z_{j,1}}, & \xi \in \hat{\Gamma}_j \\ 0, & \xi \notin \hat{\Gamma}_j \end{cases}$$

Examples of linear ($k = 1$), quadratic ($k = 2$), and cubic ($k = 3$) shape functions may be seen in Figures 4.9–4.11, respectively.

Example 4.3.2.1

We will illustrate how equations for the shape functions can be derived. We will derive the cubic shape functions, thus $k = 3$.

Assume that values of $\phi_{j,i}$, $i = 1, \ldots, 4$, are specified at evenly spaced locations on boundary element $\hat{\Gamma}_j$. Based on Equation (4.65), the form of the cubic shape functions may be written as:

$$\left[\hat{\phi}(s)\right]_j = a_{j,0} + a_{j,1}s + a_{j,2}s^2 + a_{j,3}s^3. \tag{4.67}$$

The values of $\left[\hat{\phi}(s)\right]_j$ must agree with the known values of $\phi_{j,i}$ at the nodal points $z_{j,1}, \ldots, z_{j,4}$. Thus, we have the following conditions:

$$\begin{aligned}
\left[\hat{\phi}\left(\frac{z_{j,1} - z_{j,1}}{z_{j,4} - z_{j,1}}\right)\right]_j &= a_{j,0} + a_{j,1}\left(\frac{z_{j,1} - z_{j,1}}{z_{j,4} - z_{j,1}}\right) + a_{j,2}\left(\frac{z_{j,1} - z_{j,1}}{z_{j,4} - z_{j,1}}\right)^2 \\
&\quad + a_{j,3}\left(\frac{z_{j,1} - z_{j,1}}{z_{j,4} - z_{j,1}}\right)^3 = \phi_{j,1} \\
\left[\hat{\phi}\left(\frac{z_{j,2} - z_{j,1}}{z_{j,4} - z_{j,1}}\right)\right]_j &= a_{j,0} + a_{j,1}\left(\frac{z_{j,2} - z_{j,1}}{z_{j,4} - z_{j,1}}\right) + a_{j,2}\left(\frac{z_{j,2} - z_{j,1}}{z_{j,4} - z_{j,1}}\right)^2 \\
&\quad + a_{j,3}\left(\frac{z_{j,2} - z_{j,1}}{z_{j,4} - z_{j,1}}\right)^3 = \phi_{j,2} \\
\left[\hat{\phi}\left(\frac{z_{j,3} - z_{j,1}}{z_{j,4} - z_{j,1}}\right)\right]_j &= a_{j,0} + a_{j,1}\left(\frac{z_{j,3} - z_{j,1}}{z_{j,4} - z_{j,1}}\right) + a_{j,2}\left(\frac{z_{j,3} - z_{j,1}}{z_{j,4} - z_{j,1}}\right)^2 \\
&\quad + a_{j,3}\left(\frac{z_{j,3} - z_{j,1}}{z_{j,4} - z_{j,1}}\right)^3 = \phi_{j,3} \\
\left[\hat{\phi}\left(\frac{z_{j,4} - z_{j,1}}{z_{j,4} - z_{j,1}}\right)\right]_j &= a_{j,0} + a_{j,1}\left(\frac{z_{j,4} - z_{j,1}}{z_{j,4} - z_{j,1}}\right) + a_{j,2}\left(\frac{z_{j,4} - z_{j,1}}{z_{j,4} - z_{j,1}}\right)^2 \\
&\quad + a_{j,3}\left(\frac{z_{j,4} - z_{j,1}}{z_{j,4} - z_{j,1}}\right)^3 = \phi_{j,4}.
\end{aligned}$$

Given the assumption that the nodal points are evenly spaced, the above equations simplify to the following:

$$\left[\hat{\phi}(0)\right]_j = a_{j,0} + a_{j,1}(0) + a_{j,2}(0)^2 + a_{j,3}(0)^3 = \phi_{j,1}$$
$$\left[\hat{\phi}(1/3)\right]_j = a_{j,0} + a_{j,1}(1/3) + a_{j,2}(1/3)^2 + a_{j,3}(1/3)^3 = \phi_{j,2}$$
$$\left[\hat{\phi}(2/3)\right]_j = a_{j,0} + a_{j,1}(2/3) + a_{j,2}(2/3)^2 + a_{j,3}(2/3)^3 = \phi_{j,3}$$
$$\left[\hat{\phi}(1)\right]_j = a_{j,0} + a_{j,1}(1) + a_{j,2}(1)^2 + a_{j,3}(1)^3 = \phi_{j,4}.$$

This implies the following matrix system:

$$\begin{bmatrix} 1 & 0 & 0 & 0 \\ 1 & 1/3 & 1/9 & 1/27 \\ 1 & 2/3 & 4/9 & 8/27 \\ 1 & 1 & 1 & 1 \end{bmatrix} \begin{bmatrix} a_{j,0} \\ a_{j,1} \\ a_{j,2} \\ a_{j,3} \end{bmatrix} = \begin{bmatrix} \phi_{j,1} \\ \phi_{j,2} \\ \phi_{j,3} \\ \phi_{j,4} \end{bmatrix}.$$

The solution of this system of equations is:

$$\begin{bmatrix} a_{j,0} \\ a_{j,1} \\ a_{j,2} \\ a_{j,3} \end{bmatrix} = \begin{bmatrix} \phi_{j,1} \\ -\frac{11}{2}\phi_{j,1} + 9\phi_{j,2} - \frac{9}{2}\phi_{j,3} + \phi_{j,4} \\ 9\phi_{j,1} - \frac{45}{2}\phi_{j,2} + 18\phi_{j,3} - \frac{9}{2}\phi_{j,4} \\ -\frac{9}{2}\phi_{j,1} + \frac{27}{2}\phi_{j,2} - \frac{27}{2}\phi_{j,3} + \frac{9}{2}\phi_{j,4} \end{bmatrix}.$$

Thus, Equation (4.67) becomes:

$$\begin{aligned} \left[\hat{\phi}(s)\right]_j &= a_{j,0} + a_{j,1}s + a_{j,2}s^2 + a_{j,3}s^3 \\ &= \left(\phi_{j,1}\right) + \left(-\frac{11}{2}\phi_{j,1} + 9\phi_{j,2} - \frac{9}{2}\phi_{j,3} + \phi_{j,4}\right)s \\ &\quad + \left(9\phi_{j,1} - \frac{45}{2}\phi_{j,2} + 18\phi_{j,3} - \frac{9}{2}\phi_{j,4}\right)s^2 \\ &\quad + \left(-\frac{9}{2}\phi_{j,1} + \frac{27}{2}\phi_{j,2} - \frac{27}{2}\phi_{j,3} + \frac{9}{2}\phi_{j,4}\right)s^3. \end{aligned}$$

Re-arranging these terms so that we have a linear combination of the nodal values $\phi_{j,i}$, which is the form of Equation (4.64), we get:

$$\left[\hat{\phi}(s)\right]_j = \underbrace{\left(1 - \frac{11}{2}s + 9s^2 - \frac{9}{2}s^3\right)}_{N^3_{j,1}(s)} \phi_{j,1} + \underbrace{\left(9s - \frac{45}{2}s^2 + \frac{27}{2}s^3\right)}_{N^3_{j,2}(s)} \phi_{j,2}$$
$$+ \underbrace{\left(-\frac{9}{2}s + 18s^2 - \frac{27}{2}s^3\right)}_{N^3_{j,3}(s)} \phi_{j,3} + \underbrace{\left(s - \frac{9}{2}s^2 + \frac{9}{2}s^3\right)}_{N^3_{j,4}(s)} \phi_{j,4}.$$

Upon identification with Equation (4.64), we obtain the following equations:

$$N^3_{j,1}(s) = 1 - \frac{11}{2}s + 9s^2 - \frac{9}{2}s^3$$
$$N^3_{j,2}(s) = 9s - \frac{45}{2}s^2 + \frac{27}{2}s^3$$
$$N^3_{j,3}(s) = -\frac{9}{2}s + 18s^2 - \frac{27}{2}s^3$$
$$N^3_{j,4}(s) = s - \frac{9}{2}s^2 + \frac{9}{2}s^3.$$

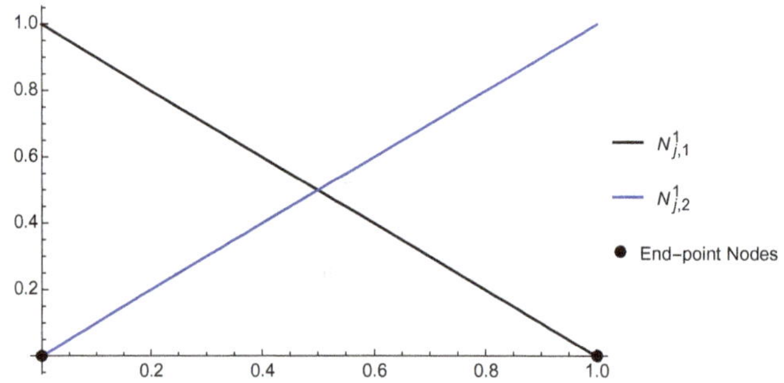

Figure 4.9: Example of linear shape functions on a 2-node boundary element. Each shape function attains a value of 1 at one of the nodal points and is 0 at all other nodal points within the boundary element. The nodal points are evenly spaced on $\hat{\Gamma}_j$. Further, each shape function is identically 0 on every other boundary element.

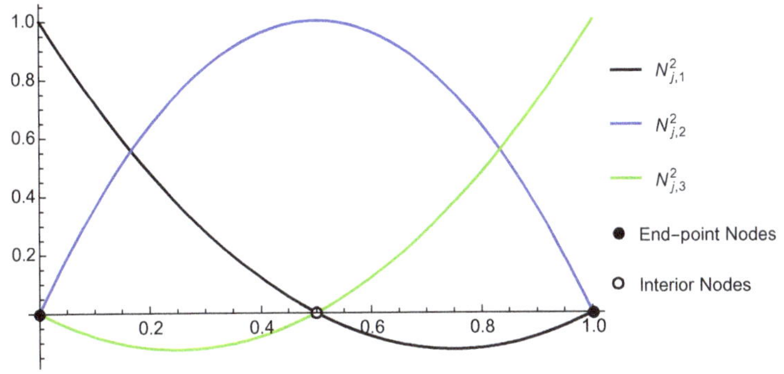

Figure 4.10: Example of quadratic shape functions on a 3-node boundary element. Each shape function attains a value of 1 at one of the nodal points and is 0 at all other nodal points within the boundary element. The nodal points are evenly spaced on $\hat{\Gamma}_j$. Further, each shape function is identically 0 on every other boundary element.

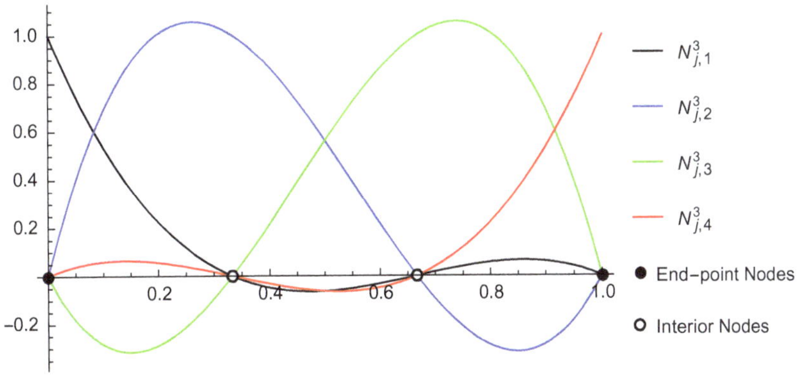

Figure 4.11: Example of cubic shape functions on a 4-node boundary element. Each shape function attains a value of 1 at one of the nodal points and is 0 at all other nodal points within the boundary element. The nodal points are evenly spaced on $\hat{\Gamma}_j$. Further, each shape function is identically 0 on every other boundary element.

4.3.3 Direct numerical integration of the Cauchy integral equation—general form

Next, we substitute the assumed forms of $\left[\hat{\phi}(s)\right]_j$ and $\left[\hat{\psi}(s)\right]_j$ into Equation (4.61). When we do this, the new approximation of $\omega(z)$ that uses the particular functions specified in Equation (4.64) is denoted $\hat{\omega}$. Thus,

$$\begin{aligned}\hat{\omega}(z) &= \frac{1}{2\pi i}\sum_{j=1}^{m}\int_0^1 \frac{\left(\sum_{i=1}^{k+1} N_{j,i}^k(s)\phi_{j,i}\right)\mathrm{d}s}{s+\chi_j(z)}\\ &+ \frac{i}{2\pi i}\sum_{j=1}^{m}\int_0^1 \frac{\left(\sum_{i=1}^{k+1} N_{j,i}^k(s)\psi_{j,i}\right)\mathrm{d}s}{s+\chi_j(z)}, \qquad z\in\hat{\Omega}.\end{aligned} \tag{4.68}$$

Substituting the polynomial forms of $\sum_{i=1}^{k+1} N_{j,i}^k(s)\phi_{j,i}$ and $\sum_{i=1}^{k+1} N_{j,i}^k(s)\psi_{j,i}$ that were given in Equation (4.65), Equation (4.68) becomes:

$$\begin{aligned}\hat{\omega}(z) &= \frac{1}{2\pi i}\sum_{j=1}^{m}\int_0^1 \frac{\left(a_{j,0}+a_{j,1}s+\cdots+a_{j,k}s^k\right)\mathrm{d}s}{s+\chi_j(z)}\\ &+ \frac{i}{2\pi i}\sum_{j=1}^{m}\int_0^1 \frac{\left(b_{j,0}+b_{j,1}s+\cdots+b_{j,k}s^k\right)\mathrm{d}s}{s+\chi_j(z)}, \qquad z\in\hat{\Omega}\\ &= \frac{1}{2\pi i}\sum_{j=1}^{m}\sum_{p=0}^{k} a_{j,p}\int_0^1 \frac{s^p\,\mathrm{d}s}{s+\chi_j(z)}\\ &+ \frac{i}{2\pi i}\sum_{j=1}^{m}\sum_{p=0}^{k} b_{j,p}\int_0^1 \frac{s^p\,\mathrm{d}s}{s+\chi_j(z)}, \qquad z\in\hat{\Omega}\\ &= \frac{1}{2\pi i}\sum_{j=1}^{m}\sum_{p=0}^{k}\left(a_{j,p}+ib_{j,p}\right)\int_0^1 \frac{s^p\,\mathrm{d}s}{s+\chi_j(z)}, \qquad z\in\hat{\Omega}.\end{aligned} \tag{4.69}$$

Thus, we see from Equation (4.69) that it is necessary to evaluate integral expressions of the form: $\int_0^1 \frac{s^p\mathrm{d}s}{s+\chi_j(z)}$, where $p=0,\ldots,k$. The necessary integrals are given below:

$$\int_0^1 \frac{s^0\,\mathrm{d}s}{s+\chi_j(z)} = \ln\left(s+\chi_j(z)\right)\Big|_0^1 = -\ln\left(\frac{\chi_j(z)}{1+\chi_j(z)}\right)$$

$$\int_0^1 \frac{s^1\,\mathrm{d}s}{s+\chi_j(z)} = s-\chi_j(z)\ln\left(s+\chi_j(z)\right)\Big|_0^1$$

$$= 1+\chi_j(z)\ln\left(\frac{\chi_j(z)}{1+\chi_j(z)}\right)$$

$$\int_0^1 \frac{s^2\,\mathrm{d}s}{s+\chi_j(z)} = \frac{1}{2}s^2-\chi_j(z)s+\left(\chi_j(z)\right)^2\ln\left(s+\chi_j(z)\right)\Big|_0^1$$

$$= \frac{1}{2}-\chi_j(z)-\left(\chi_j(z)\right)^2\ln\left(\frac{\chi_j(z)}{1+\chi_j(z)}\right)$$

$$\vdots$$

$$\int_0^1 \frac{s^p\,\mathrm{d}s}{s+\chi_j(z)} = \sum_{i=0}^{p-1}\left(\frac{\left(-\chi_j(z)\right)^i s^{p-i}}{p-i}\right)+\left(-\chi_j(z)\right)^p\ln\left(s+\chi_j(z)\right)\Big|_0^1$$

$$= \sum_{i=0}^{p-1}\left(\frac{\left(-\chi_j(z)\right)^i}{p-i}\right)-\left(-\chi_j(z)\right)^p\ln\left(\frac{\chi_j(z)}{1+\chi_j(z)}\right).$$

Further, by algebraic simplification, we obtain the following identity:

$$\ln\left(\frac{\chi_j(z)}{1+\chi_j(z)}\right) = \ln\left(\frac{\frac{z_j-z}{z_{j+1}-z_j}}{1+\frac{z_j-z}{z_{j+1}-z_j}}\right)$$

$$= \ln\left(\frac{\frac{z_j-z}{z_{j+1}-z_j}}{1+\frac{z_j-z}{z_{j+1}-z_j}}\times\frac{z_{j+1}-z_j}{z_{j+1}-z_j}\right)$$

$$= \ln\left(\frac{z_j-z}{z_{j+1}-z_j+z_j-z}\right)$$

$$= \ln\left(\frac{z_j-z}{z_{j+1}-z}\right)$$

$$= \ln\left(\frac{z_j-z}{z_{j+1}-z}\times\frac{-1}{-1}\right)$$

$$= \ln\left(\frac{z-z_j}{z-z_{j+1}}\right).$$

Note 4.3.3.1

Hence, we conclude the following CVBEM approximation function:

$$\begin{aligned}\hat{\omega}(z) &= \frac{1}{2\pi i}\sum_{j=1}^{m}\sum_{p=0}^{k}(a_{j,p}+ib_{j,p})\int_{0}^{1}\frac{s^{p}\,\mathrm{d}s}{s+\chi_j(z)}, \qquad z\in\hat{\Omega} \\ &= \frac{1}{2\pi i}\sum_{j=1}^{m}\sum_{p=0}^{k}c_{j,p}\int_{0}^{1}\frac{s^{p}\,\mathrm{d}s}{s+\chi_j(z)}, \qquad z\in\hat{\Omega}\end{aligned} \tag{4.70}$$

where the definite integral may be computed directly by:

$$\int_{0}^{1}\frac{s^{p}\,\mathrm{d}s}{s+\chi_j(z)} = \sum_{i=0}^{p-1}\left(\frac{\left(-\chi_j(z)\right)^{i}}{p-i}\right) - \left(-\chi_j(z)\right)^{p}\ln\left(\frac{z-z_j}{z-z_{j+1}}\right). \tag{4.71}$$

4.3.4 Direct numerical integration of the Cauchy integral equation—linear shape functions

In this section, we are going to consider a very important case of Equations (4.70) and (4.71). Namely, we will consider the case in which linear shape functions are used to approximate the boundary conditions on each Γ_j in the development of the CVBEM approximation function. That is, only element end-point nodes—and not element interior nodes—are specified on each Γ_j. Therefore, each $\hat{\Gamma}_j$ resembles Figure 4.12.

Moreover, in this case, the approximate polygonal boundary, $\hat{\Gamma}_j$, will resemble Figure 4.13.

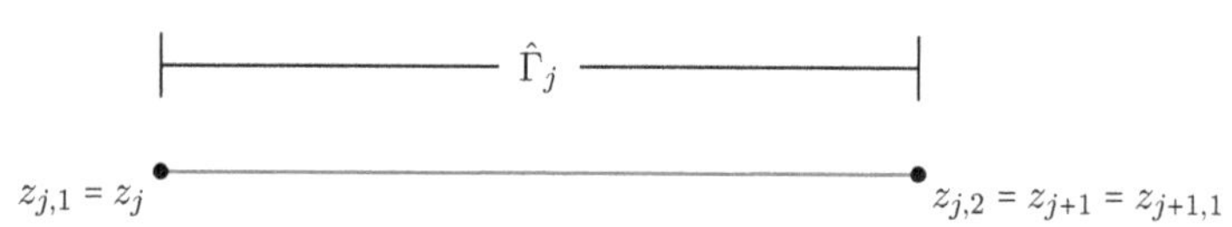

Figure 4.12: A boundary element that only incorporates the use of element end-point nodes. In this section, we will not use element interior nodes.

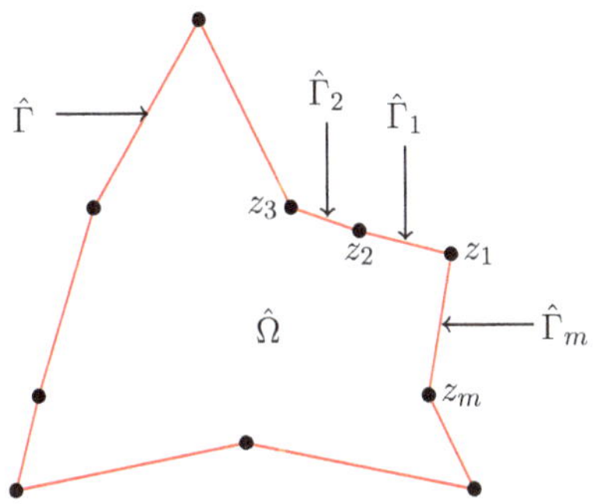

Figure 4.13: The polygonal approximation of the target problem boundary. The labeling scheme for the boundary nodal points is shown as well as the labeling scheme for the line segments $\hat{\Gamma}_1, \ldots, \hat{\Gamma}_2$. Note that only element end-point nodes are considered. In the special case that is examined in this section, we assume that no interior nodal points are used in the development of the CVBEM basis functions.

This special case was considered in [21] and [1] and is often used in numerical implementations of the CVBEM. Additionally, recall that linear shape functions were depicted in Figure 4.9. The corresponding equations for the linear shape functions are as follows:

$$\begin{aligned} N_{j,1}^1(s) &= 1 - s, && 0 \le s \le 1 \\ N_{j,2}^1(s) &= s, && 0 \le s \le 1 \end{aligned}$$

Thus, we may model the CVBEM approximation function on $\hat{\Gamma}_j$ as follows:

$$\begin{aligned} \left[\hat{\omega}(s)\right]_j &= N_{j,1}^1(s)\omega_j + N_{j,2}^1(s)\omega_{j+1}, && 0 \le s \le 1 \\ &= (1 - s)\omega_j + (s)\omega_{j+1}, && 0 \le s \le 1 \\ &= (\omega_{j+1} - \omega_j)s + \omega_j, && 0 \le s \le 1. \end{aligned} \tag{4.72}$$

Now, recall $\omega_j = \phi_j + i\psi_j$. Thus, Equation (4.72), becomes:

$$\left[\hat{\omega}(s)\right]_j = (\phi_{j+1} - \phi_j)s + \phi_j + i\left[(\psi_{j+1} - \psi_j)s + \psi_j\right], \quad 0 \le s \le 1 \tag{4.73}$$

We identify the results of Equation (4.73) with Equation (4.64) as follows:

$$\begin{aligned} \left[\hat{\phi}(s)\right]_j &= \phi_j + (\phi_{j+1} - \phi_j)s \\ \left[\hat{\psi}(s)\right]_j &= \psi_j + (\psi_{j+1} - \psi_j)s. \end{aligned}$$

Thus, by identification with Equation (4.65), we have:

$$a_{j,0} = \phi_j, \qquad a_{j,1} = \phi_{j+1} - \phi_j,$$
$$b_{j,0} = \psi_j, \qquad b_{j,1} = \psi_{j+1} - \psi_j.$$

Equivalently,

$$\begin{aligned} c_{j,0} &= a_{j,0} + ib_{j,0} = \phi_j + i\psi_j = \omega_j \\ c_{j,1} &= a_{j,1} + ib_{j,1} = \phi_{j+1} - \phi_j + i(\psi_{j+1} - \psi_j) = \omega_{j+1} - \omega_j. \end{aligned} \tag{4.74}$$

The task now is to calculate the contribution of each boundary element to the global CVBEM approximation function, $\hat{\omega}(z) = \hat{\phi}(z) + i\hat{\psi}(z)$. The contribution of each $\hat{\Gamma}_j$ to $\hat{\omega}(z)$ can be computed using Equations (4.70) and (4.71) as follows:

$$\begin{aligned}
\hat{\omega}(z) &= \frac{1}{2\pi i}\sum_{j=1}^{m}\left[\omega_j \int_0^1 \frac{s^0\,\mathrm{d}s}{s+\chi_j(z)} + (\omega_{j+1}-\omega_j)\int_0^1 \frac{s^1\,\mathrm{d}s}{s+\chi_j(z)}\right]_j, && z \in \hat{\Omega} \\
&= \frac{1}{2\pi i}\sum_{j=1}^{m}\left[-\omega_j \ln\left(\frac{z-z_j}{z-z_{j+1}}\right)\right. \\
&\qquad \left. + (\omega_{j+1}-\omega_j)\left(1+\left(\frac{z_j - z}{z_{j+1}-z_j}\right)\ln\left(\frac{z-z_j}{z-z_{j+1}}\right)\right)\right], && z \in \hat{\Omega} \\
&= \frac{1}{2\pi i}\sum_{j=1}^{m}\left[-\omega_j \ln\left(\frac{z-z_j}{z-z_{j+1}}\right) + (\omega_{j+1}-\omega_j)\right. \\
&\qquad + \omega_{j+1}\left(\frac{z_j - z}{z_{j+1}-z_j}\right)\ln\left(\frac{z-z_j}{z-z_{j+1}}\right) \\
&\qquad \left. - \omega_j\left(\frac{z_j - z}{z_{j+1}-z_j}\right)\ln\left(\frac{z-z_j}{z-z_{j+1}}\right)\right]_j, && z \in \hat{\Omega} \\
&= \frac{1}{2\pi i}\sum_{j=1}^{m}\left\{\omega_{j+1}\left[1+\left(\frac{z_j - z}{z_{j+1}-z_j}\right)\ln\left(\frac{z-z_j}{z-z_{j+1}}\right)\right]\right. \\
&\qquad \left. - \omega_j\left[1+\ln\left(\frac{z-z_j}{z-z_{j+1}}\right)+\left(\frac{z_j - z}{z_{j+1}-z_j}\right)\ln\left(\frac{z-z_j}{z-z_{j+1}}\right)\right]\right\}, && z \in \hat{\Omega} \\
&= \frac{1}{2\pi i}\sum_{j=1}^{m}\left\{\omega_{j+1}\left[1+\left(\frac{z_j - z}{z_{j+1}-z_j}\right)\ln\left(\frac{z-z_j}{z-z_{j+1}}\right)\right]\right. \\
&\qquad \left. - \omega_j\left[1+\left(\frac{z_{j+1} - z}{z_{j+1}-z_j}\right)\ln\left(\frac{z-z_j}{z-z_{j+1}}\right)\right]\right\}, && z \in \hat{\Omega}.
\end{aligned}$$

Note 4.3.4.1

When linear shape functions are used, the CVBEM approximation function takes the following form:

$$\hat{\omega}(z) = \frac{1}{2\pi i}\sum_{j=1}^{m}\left\{\omega_{j+1}\left[1+\left(\frac{z_j - z}{z_{j+1}-z_j}\right)\ln\left(\frac{z-z_j}{z-z_{j+1}}\right)\right]\right.$$
$$\left. - \omega_j\left[1+\left(\frac{z_{j+1} - z}{z_{j+1}-z_j}\right)\ln\left(\frac{z-z_j}{z-z_{j+1}}\right)\right]\right\}, \qquad z \in \hat{\Omega}.$$

Finally, recall the expression of the CVBEM approximation function given in Equation (4.63) with $k = 1$:

$$\begin{aligned} \hat{\omega}(z) &= \frac{1}{2\pi i}\sum_{j=1}^{m}\int_0^1 \frac{\left[\hat{\omega}(s)\right]_j \mathrm{d}s}{s+\chi_j(z)}, & z \in \hat{\Omega} \\ &= \frac{1}{2\pi i}\sum_{j=1}^{m}\int_0^1 \frac{\sum_{i=1}^{2} N_{j,i}^1(s)\omega_{j,i}\,\mathrm{d}s}{s+\chi_j(z)}, & z \in \hat{\Omega} \\ &= \frac{1}{2\pi i}\sum_{j=1}^{m}\sum_{i=1}^{2}\int_0^1 \frac{N_{j,i}^1(s)\omega_{j,i}\,\mathrm{d}s}{s+\chi_j(z)}, & z \in \hat{\Omega}. \end{aligned} \tag{4.75}$$

4.3.5 Additional CVBEM basis functions

An important consideration when modeling with the CVBEM is to choose basis functions that are analytic within the problem domain. This is necessary so that linear combinations of the basis functions, and specifically, the CVBEM approximation function itself, are analytic within the problem domain. By using such analytic functions, the CVBEM approximation function has the following properties:

1. The CVBEM approximation functions provides continuous computational estimates throughout the problem domain.
2. Both the real and imaginary parts of the CVBEM approximation function satisfy the two-dimensional Laplace equation throughout the problem domain.

These properties of CVBEM approximation functions distinguish the CVBEM from real-variable domain discretization techniques such as the finite element (FEM) and finite difference method (FDM), which develop estimates only at nodal points defined by meshing the problem domain. First, since the FEM and FDM models only result in estimates of the solution at a finite number of points in the domain, it is necessary to use some type of interpolation to obtain estimates of the solution at any other point in the domain other than a mesh point. Second, methods such as the FEM are designed to solve the weak formulation of the governing PDE. However, as a consequence of the Cauchy–Riemann equations, both the real and imaginary components of the CVBEM approximation function solve the strong formulation of Laplace's equation.

In the previous sections, we dedicated a significant amount of space to the development of the basis functions resulting from the numerical integration of the Cauchy integral equation using linear shape functions on the boundary. These basis functions yield a CVBEM approximation function of the following form:

$$\hat{\omega}(z) = \sum_{j=1}^{n} c_j(z - z_j)\ln(z - z_j), \quad c_j \in \mathbb{C}. \tag{4.76}$$

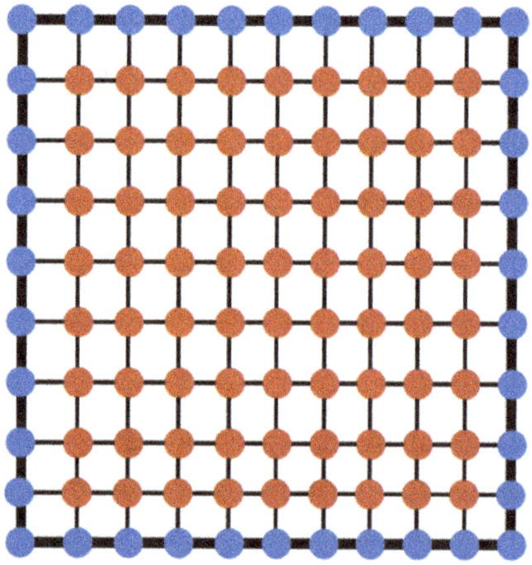

Figure 4.14: Example of a finite element mesh of a rectangular problem domain. Boundary conditions would be specified along the boundary at the blue points, and the FEM model would produce estimates of the solution at the red points. If an estimate of the solution were desired somewhere in the interior of the domain, but not at an already-existing red nodal point, it would be necessary to perform some type of interpolation to produce an estimate at the desired location.

In this section, we introduce three other types of basis functions that may be useful to the reader when constructing a CVBEM model. The three basis function families considered lead to CVBEM approximation functions of the following forms:

1. Complex polynomials, which lead to a power series approximation expanded about the point $z_0 \in \mathbb{C}$:

$$\hat{\omega}(z) = \sum_{j=1}^{n} c_j (z - z_0)^j, \quad c_j \in \mathbb{C}; \tag{4.77}$$

2. Laurent series expansion (reciprocals of complex monomials evaluated at a target expansion point, $z_1 \in \mathbb{C}$):

$$\hat{\omega}(z) = \sum_{j=1}^{n} c_j \frac{1}{(z - z_1)^j}, \quad c_j \in \mathbb{C}; \tag{4.78}$$

3. Reciprocals of complex linear polynomials evaluated at a set of nodal points:

$$\hat{\omega}(z) = \sum_{j=1}^{n} c_j \frac{1}{z - z_j}, \quad c_j \in \mathbb{C}. \tag{4.79}$$

Note 4.3.5.1

In Equations (4.76)–(4.79), z is a complex variable, z_j is a computational node, and z_0 is any point in the complex plane, and z_1 is a point in the complex plane that is located exterior to the problem domain union boundary. The basis functions given in Equation (4.78) are simple poles, and they are used in the next chapter to create a CVBEM model in *Excel*.

Suppose Dirichlet boundary data are specified on a problem domain. The CVBEM methodology can be used to determine an approximation of the potential function corresponding to the given Dirichlet boundary data. We denote the approximate potential function $\hat{\phi}$, and it is related to the analytic CVBEM approximation function by $\hat{\phi} = \text{Re}\,(\hat{\omega})$.

An advantage of using the CVBEM methodology is that the corresponding stream function, denoted $\hat{\psi}$, can be determined up to an additive constant directly from the analytic CVBEM approximation function: $\hat{\psi} = \text{Im}\,(\hat{\omega})$. This useful result is a consequence of the fact that the potential and stream functions of the target solution, as well as the CVBEM approximation function, are both conjugate functions. Numerically, this can be seen in Equation (4.90), which makes it clear that $\hat{\phi}$ and $\hat{\psi}$ are defined by the same coefficients, α_j and β_j for $j = 1, \ldots, n$.

Note 4.3.5.2

Recall the procedure for determining the corresponding stream function given a particular potential function:

$$\hat{\psi} = -\int \frac{\partial \hat{\phi}}{y}\, \mathrm{d}x + g(y), \tag{4.80}$$

where $g(y)$ is determined by:

$$g(y) = \int \left[\frac{\partial \hat{\phi}}{\partial x} + \int \frac{\partial^2 \hat{\phi}}{y^2}\, \mathrm{d}x \right] \mathrm{d}y\,. \tag{4.81}$$

Equivalently, combining Equations (4.80) and (4.81):

$$\hat{\psi} = -\int \frac{\partial \hat{\phi}}{y}\, \mathrm{d}x + \int \left[\frac{\partial \hat{\phi}}{\partial x} + \int \frac{\partial^2 \hat{\phi}}{y^2}\, \mathrm{d}x \right] \mathrm{d}y\,.$$

For more details, see Example 2.3.3.2 in Section 2.3.

The basis functions listed in Equations (4.77)–(4.79) have been previously examined, with respect to modeling potential flow in a 90-degree bend, in [22]. Another comparison study was done in [23]. Sample results from the analysis of [22] are depicted in Figure 4.15. Since both the real and imaginary parts of the approximation function are conjugate functions, their contours are orthogonal where they intersect. When plotted, these orthogonal contours produce the standard flow net graphical display, which is elaborated upon in [24]. The following figures make use of the flow net graphical representation of potential functions.

For the purpose of demonstrating the fidelity of the various basis function families, the problem under consideration is modeling the potential flow of an ideal fluid around a 90-degree bend. The analytic function describing such ideal fluid flow is $\omega(z) = z^2$, with real and imaginary parts as follows:

$$\phi(x, y) = \mathrm{Re}\left[\omega(z)\right] = x^2 - y^2 \quad \text{and} \quad \mathrm{Im}\left[\omega(z)\right] = 2xy. \tag{4.82}$$

The purpose of using a test problem with a known solution—such as the problem described in Equation (4.82)—is that we will be able to assess the accuracy of the CVBEM approximation functions by direct comparison with the known solution.

Note 4.3.5.3

For this particular test problem, notice that the exact solution, $\omega(z) = z^2$, is contained within the basis function family of the complex polynomials $(z - z_0)^j$ in the case of $z_0 = 0$ and $j = 2$. That is, the exact solution can be obtained as a linear combination of the polynomial basis functions. However, while such a situation would be desirable in real-world modeling, we wish to assess the ability of the complex polynomials to model a function that is not in the span of their basis functions. Consequently, for the purpose of this demonstration, we have removed $(z - z_0)^2$ from the basis of the complex polynomial family of functions. Thus, the complex polynomial model employed for this analysis has the form:

$$\hat{\omega}(z) = (\alpha_1 + i\beta_1)(z - z_0) + \sum_{k=2}^{n}(\alpha_k + i\beta_k)(z - z_0)^{k+1}.$$

The major conclusion of this section is that there are a number of suitable basis function families for use with CVBEM modeling. In fact, any function that is analytic within the problem domain can be used as a basis function in the development of a CVBEM approximation function. Additionally, the basis functions of a CVBEM approximation function can include members from various different basis function families.

Example 4.3.5.1

An example of a CVBEM approximation function using members from different basis function families might be as follows:

$$\hat{\omega}(z) = \sum_{j=1}^{4} c_{j,1}(z - z_j)\ln(z - z_j) + \sum_{j=1}^{5} c_{j,2}(z - z_0)^j + \sum_{j=1}^{6} c_{j,3}\frac{1}{z - z_j}. \tag{4.83}$$

Of course, it is necessary for the z_j in Equation (4.83) to be located outside of the problem domain so that each of the basis functions used in this approximation function are analytic within the problem domain.

Consequently, since each basis function used in this CVBEM approximation function is analytic within the problem domain, it follows that the CVBEM approximation function itself is analytic within the problem domain because the sum of analytic functions is an analytic function, as demonstrated in Section 2.3.

In this example, the CVBEM approximation function uses members from three different types of basis function families. A total of 15 basis functions are used, which means there are a total of 30 unknown coefficients to be determined. These coefficients are denoted $c_{j,1}$, $c_{j,2}$, and $c_{j,3}$, where the numbers 1, 2, and 3 are used to distinguish each of the three different basis function families.

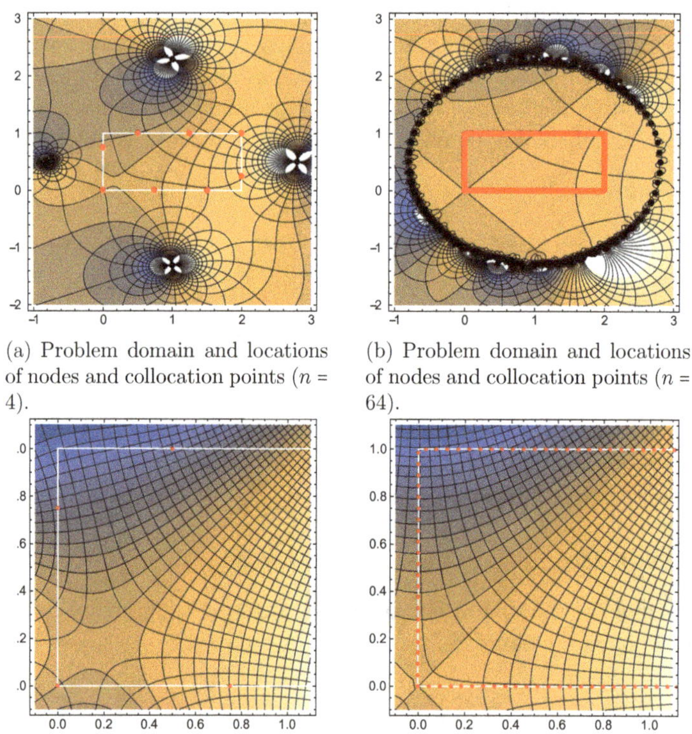

(a) Problem domain and locations of nodes and collocation points (n = 4).

(b) Problem domain and locations of nodes and collocation points (n = 64).

(c) Approximation of potential flow in a 90-degree bend (n = 4).

(d) Approximation of potential flow in a 90-degree bend (n = 64).

Figure 4.15: Demonstration of a CVBEM model of potential flow in a 90-degree bend using the simple pole basis functions $1/(z - z_j)$.

The secondary conclusion of this section is the ease with which it is possible to develop flow nets using the CVBEM methodology. Given Dirichlet boundary conditions from the target potential function, the CVBEM methodology results in an approximation of both the potential function and the associated stream function corresponding to the given boundary data.

Note 4.3.5.4

It should be emphasized that the CVBEM methodology does not require the use of another software program or post-processing routine to generate the orthogonal streamlines that are depicted in Figure 4.15. Rather, the orthogonal contours that are depicted in the flow nets are a direct outcome of the CVBEM itself. This property is among the several properties that distinguish the CVBEM from the FDM and FEM approximations, which require a post-processing procedure to produce estimates of the stream function associated with the computed potential function.

The choice of the basis function families that were discussed in this section was arbitrary, with the condition that the basis functions must be analytic over at least the domain of interest. However, the selections that were made follow the construct seen in the commonly used series expansions such as the Taylor series and Laurent series. Additionally, we examined basis functions resembling the idea of distributed source terms as well as the family of basis functions resulting from the numerical integration of the Cauchy integral equation. As previously mentioned, other basis function families can readily be substituted for these in a CVBEM approximation function.

Of course, the particular choice of basis functions used in a CVBEM approximation function can lead to better approximation performance for some types of boundary value problems and poorer approximation performance for other problems. Nonetheless, regardless of the basis functions used in the model, the resulting CVBEM approximation function will be analytic over the problem domain and, hence, will have all of the various properties associated with analytic functions.

4.4 Numerical implementation of the CVBEM

In this section:

1. Using matrix equations to determine the coefficients of the CVBEM approximation function.

2. Handling the rotation of branch cuts.

3. Using the computed coefficients of $\hat{\phi}$ to obtain $\hat{\psi}$.

The CVBEM is a generalized boundary integral equation method based on the Cauchy integral theorem for analytic complex variable functions [3]. While the CVBEM was originally conceived to approximate the solution to two-dimensional Laplace boundary value problems, the method has since evolved to be capable of approximation of Laplace boundary value problems in three and higher spatial dimensions [25]. Furthermore, the CVBEM has also been expanded over the years to include capabilities of modeling both Dirichlet and mixed boundary value problems [24]. Due to the CVBEM's ability to model a variety of problems of interest in science and engineering, as well as the CVBEM's use of computational nodes when using the basis functions of Equation (4.87), it is a useful PDE numerical method for demonstrating the proposed NPA. The remainder of this section develops the CVBEM basis functions and illuminates the relationship between the CVBEM basis functions and the computational nodes that are the focus of the NPA.

Let $\Omega \subseteq \mathbb{C}$ denote a simply connected problem domain, and let $\omega : \Omega \to \mathbb{C}$ be an analytic function of the complex variable $z = x + iy$, where $i = \sqrt{-1}$ and $x, y \in \mathbb{R}$. The complex variable function ω consists of two real variable functions, denoted $\phi : \mathbb{R}^2 \to \mathbb{R}$ and $\psi : \mathbb{R}^2 \to \mathbb{R}$, respectively, such that $\omega(z) = \phi(x,y) + i\psi(x,y)$. Since ω is analytic, ϕ and ψ are related by the Cauchy–Riemann equations:

$$\frac{\partial \phi}{\partial x} = \frac{\partial \psi}{\partial y} \quad \text{and} \quad \frac{\partial \phi}{\partial y} = -\frac{\partial \psi}{\partial x}. \tag{4.84}$$

A consequence of Equation (4.84) is that both ϕ and ψ are harmonic functions. The key insight of the CVBEM approach is to create an approximation function, denoted $\hat{\omega}$, consisting of a linear combination of complex variable functions that are each analytic within Ω so that both the real and imaginary parts of $\hat{\omega}$ satisfy Laplace's equation within Ω:

$$\begin{aligned} \Delta\hat{\phi} &= 0, \qquad \text{where } \hat{\phi} = \mathrm{Re}\left[\hat{\omega}(z)\right] \\ \Delta\hat{\psi} &= 0, \qquad \text{where } \hat{\psi} = \mathrm{Im}\left[\hat{\omega}(z)\right] \end{aligned} \tag{4.85}$$

The $g_j(z) : \mathbb{C} \to \mathbb{C}$ are analytic within Ω and are selected by the modeler. The coefficients $c_j \in \mathbb{C}$ are complex numbers, each composed of two real constants, $\mathrm{Re}(c_j)$ and $\mathrm{Im}(c_j)$. Since each c_j corresponds to two degrees of freedom, there are a total of $2n$ degrees of freedom to be determined in a CVBEM model. The particular values of these degrees of freedom are dependent on the geometry of the problem, the boundary conditions, the selected $g_j(z)$, and the locations of the computational nodes as determined by the NPA.

Since the real and imaginary parts of $\hat{\omega}$ are harmonic functions within Ω, the CVBEM approximation function satisfies the governing PDE within the problem domain. Hence, the only error in Equation (4.53) is due to error in continuously satisfying the boundary conditions. Therefore, the $2n$ degrees of freedom are chosen so that the CVBEM approximation function satisfies the given boundary conditions as closely as possible within a given norm. That is, the coefficients of the CVBEM approximation function are determined so that $\hat{\omega}$ approximates the analytic solution ω. Note: in this section, "degrees of freedom" is used exclusively to refer to the $2n$ unknown values of the coefficients in Equation (4.53) and not to the locations of the candidate nodes or candidate collocation points for the NPA.

While there have been CVBEM implementations using different basis functions (see [22]), only certain analytic functions require the use of computational nodes. In order to demonstrate the NPA, we will use analytic functions obtained by direct numerical integration of the Cauchy integral equation, as these basis functions incorporate the use of computational nodes. The Cauchy integral equation is

$$f(z_0) = \frac{1}{2\pi i} \oint_{\partial\Omega} \frac{f(z)dz}{z - z_0}, \tag{4.86}$$

and as demonstrated in [3], when straight line segments are used to discretize the boundary of the problem domain, the numerical integration of Equation (4.86) results in the following sum:

$$\hat{\omega}(z) = \sum_{j=1}^{n} c_j (z - z_j) \ln(z - z_j). \tag{4.87}$$

The $z_j \in \mathbb{C}$ of Equation (4.87) are branch points of the functions $(z - z_j)\ln(z - z_j)$. Importantly, within the context of the NPA, these branch points are the computational nodes whose locations within $\mathcal{N}$ will be determined by the NPA. These nodes are referred to as "computational nodes" because they do not have a physical meaning within the problem context and they only exist due to the use of these particular basis functions. Further, their locations can be selected arbitrarily within $\mathcal{N}$ (as long as each node is in a distinct location, which is necessary for the basis functions to be linearly independent) and there will not be a change in the physical interpretation of the approximation function, rather, the only changes to the CVBEM model will be in the values of the coefficients and the accuracy of the resulting approximation function.

Each basis function, $(z - z_j)\ln(z - z_j)$, in the linear combination of Equation (4.87) has a branch cut emanating from z_j. The CVBEM approximation function is discontinuous along each branch cut. Therefore, in order to obtain a CVBEM approximation function that is continuous within the problem domain, it is necessary to rotate the branch cut associated with each basis function away from the problem domain, as illustrated in Figure 4.16. The branch cuts should be rotated such that no branch cut intersects the problem domain. Therefore, when using the NPA in conjunction with the CVBEM, it is necessary to determine not only the location of the candidate nodes, but also the rotation angle of the corresponding branch cut that prevents the branch cut from intersecting the problem domain.

Rotating the branch cuts is most conveniently accomplished in polar coordinates. Using the usual conversion to polar coordinates, let $z - z_j = R_j e^{i\theta_j}$. Then, the basis functions are represented in polar coordinates as follows:

$$\begin{aligned}(z - z_j)\ln(z - z_j) &= R_j\left[\cos(\theta_j)\ln(R_j) - \theta_j\sin(\theta_j)\right] \\ &+ iR_j\left[\theta_j\cos(\theta_j) + \sin(\theta_j)\ln(R_j)\right].\end{aligned} \tag{4.88}$$

The branch cuts are rotated by adding $\Delta\theta_j$ to the angle θ_j, where $\Delta\theta_j$ is the clockwise rotation angle for the jth basis function, see Figure 4.16. The resulting angle, denoted θ'_j, is thus $\theta'_j = \theta_j + \Delta\theta_j$. The rotation angle $\Delta\theta_j$ can be selected by the modeler as long as the rotated branch cut does not intersect the problem domain. Additionally, our experience suggests selecting a rotation angle for each branch cut such that the resulting orientation of the branch cut does not intersect any of the other branch cuts. The basis functions with rotated branch cuts are given by $R_j\left[\cos\left(\theta'_j\right)\ln\left(R_j\right) - \theta'_j\sin\left(\theta'_j\right)\right] + iR_j\left[\theta'_j\cos\left(\theta'_j\right) + \sin\left(\theta'_j\right)\ln\left(R_j\right)\right]$.

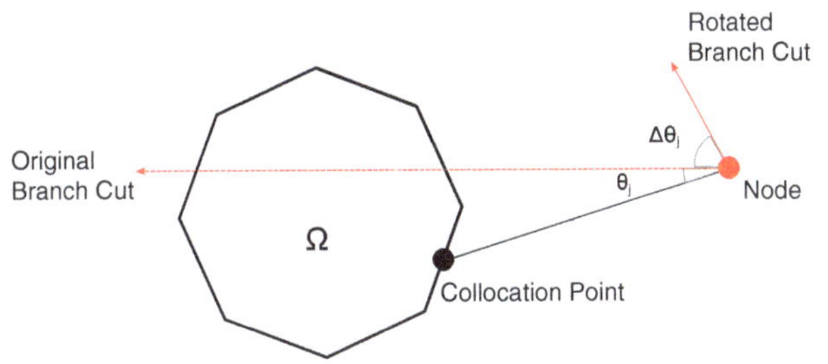

Figure 4.16: Illustration of the rotation of a typical branch cut. Some angle $\Delta\theta_j$ is added to θ_j so that the branch cut corresponding to $(z - z_j)\ln(z - z_j)$ does not intersect the problem domain or other branch cuts.

The ability to directly integrate Equation (4.86), which results in an approximation of the integral equation as a finite sum, means that there is no need to do numerical quadrature in the computer implementation of the CVBEM procedure. Instead, the only computational task after the locations of the collocation points and computational nodes have been selected by the NPA is to compute the coefficients of the linear combination of the CVBEM approximation function. These coefficients are complex and hence have both a real and an imaginary part: $c_j = \alpha_j + i\beta_j$. Likewise, the basis functions in Equation (4.53) are complex variable functions and hence also have both a real and an imaginary part: $g_j(z) = \lambda_j(x,y) + i\mu_j(x,y)$. Thus, Equation (4.53) is equivalent to

$$\begin{aligned}\hat{\omega}(z) &= \sum_{j=1}^{n} (\alpha_j + i\beta_j)(\lambda_j(x,y) + i\mu_j(x,y)) \\ &= \sum_{j=1}^{n} \Big[\alpha_j\lambda_j(x,y) - \beta_j\mu_j(x,y) + i\alpha_j\mu_j(x,y) \\ &\qquad + i\beta_j\lambda_j(x,y)\Big].\end{aligned} \tag{4.89}$$

The real and imaginary parts of Equation (4.89) are, respectively,

$$\begin{aligned}\text{Re}\left[\hat{\omega}(z)\right] &= \hat{\phi}(x,y) = \sum_{j=1}^{n} \alpha_j\lambda_j(x,y) - \beta_j\mu_j(x,y) \\ \text{Im}\left[\hat{\omega}(z)\right] &= \hat{\psi}(x,y) = \sum_{j=1}^{n} \alpha_j\mu_j(x,y) + \beta_j\lambda_j(x,y).\end{aligned} \tag{4.90}$$

In many applications, $\hat{\phi}$ and $\hat{\psi}$ are interpreted physically as the approximate potential and stream functions, respectively. Since $\hat{\omega}$ is analytic within Ω, $\Delta\hat{\phi} = 0$ and $\Delta\hat{\psi} = 0$ in Ω.

Equation (4.90) makes it clear that the approximate stream function is immediately known as soon as the coefficients of the approximate potential function have been computed (and vice versa) since they share the same degrees of freedom, namely, the α_j and β_j. This is also a consequence of the Cauchy–Riemann equations, which can be used to determine the conjugate harmonic function (up to an additive constant) of any harmonic function. Thus, the CVBEM can be used to obtain the approximate stream function using boundary data exclusively from the potential function, which is one of the useful properties of the CVBEM.

Collocation leads to a system of $2n$ equations in $2n$ unknowns. For Dirichlet boundary value problems, the unknown coefficients are determined by collocation of either the real or imaginary part of the CVBEM approximation function (depending on the nature of the given boundary conditions) with the given boundary data. In the example Dirichlet BVP that follows, we assume boundary conditions are specified from the stream function, ψ. Consequently, we perform collocation with the imaginary part of the CVBEM approximation function. The set of $2n$ equations is given as follows. Let $i = 1, \ldots, 2n$. Then, for a boundary condition $\psi(x_i, y_i)$ specified at (x_i, y_i), the ith equation is given by

$$\begin{aligned}\hat{\psi}(x_i, y_i) &= \sum_{j=1}^{n} \alpha_j \mu_j(x_i, y_i) + \beta_j \lambda_j(x_i, y_i) \\ &= \psi(x_i, y_i).\end{aligned} \tag{4.91}$$

Using collocation to determine the coefficients of the CVBEM approximation function guarantees that the CVBEM approximation function will be exact at the locations of the $2n$ collocation points on the problem boundary.

Since $\hat{\omega}$ is a linear combination of functions that are each analytic within Ω, both $\hat{\phi}$ and $\hat{\psi}$ can be evaluated continuously within Ω. Consequently, computational estimates of both the approximate potential and stream functions are provided continuously within Ω without the need for any interpolation of either $\hat{\phi}$ or $\hat{\psi}$.

An important property of the CVBEM is that once either $\hat{\phi}$ or $\hat{\psi}$ has been found, the other can be determined uniquely up to an additive constant since $\hat{\phi}$ and $\hat{\psi}$ are conjugate harmonic functions as a result of $\hat{\omega}$ being analytic within Ω. Since $\hat{\psi}$ and $\hat{\phi}$ are harmonic conjugate functions, their contour lines (level curves), which are the stream and potential lines, respectively, are orthogonal at their points of intersection throughout the domain. When these contour lines are superimposed, the resulting plot is the standard flow net graphical display, which is a commonly used tool for visualizing potential flow in engineering and the applied sciences [24, 26].

Collocation points were spaced evenly along the boundary with the first collocation point being located at the origin and the other collocation points being placed equidistant around the boundary in a counterclockwise direction. Model nodes are positioned equally spaced along a circle of radius 1.8 that was centered at the center of the rectangle $(1, 0.5)$. All nodes are located exterior to the problem domain union boundary. Additionally, when the CVBEM basis functions of Equation (4.76) were used, the respective branch cuts were rotated so as to be directed 180 degrees away from the directed line segment from each node to the center of the rectangular domain. The first node was located at $(-0.8, 0.5)$ and all subsequent nodes were equally spaced along the described circle.

For each family of basis functions, the CVBEM approximation functions were generated by using only the position and value of each collocation point, as well as the position and branch cut rotation of each node for the basis function families of Equations (4.76) and (4.79). As with any technique for solving boundary value problems, the CVBEM operates only on the boundary values of the problem. In this research, only boundary values from the potential function were used as the collocation points. So, the streamlines in Section 4.3.5 were generated without collocating the CVBEM approximation function with any streamline boundary values. That is, they are a direct product of the CVBEM technique itself. The ability to accurately approximate streamlines is essential to modeling ideal fluid flow.

The final step is to solve the matrix system. Once these values are known, they can be substituted back into Equation (4.55) and can then be used to approximate all of the potential values within the problem domain. Likewise, these values can be substituted back into Equation (4.56) and can be used to approximate all of the streamline values within the problem domain. Note that it is possible to approximate all of the streamline values within the problem domain without knowing any streamline boundary conditions. That is, the equation for the streamline function is a direct product of the CVBEM.

In this research, boundary conditions of the potential function are specified at the collocation points. Thus, for $2n$ collocation points, the necessary CVBEM approximation functions are

$$\hat{\phi}(z_1) = \alpha_1\lambda_1(z_1) - \beta_1\mu_1(z_1) + \alpha_2\lambda_2(z_1) - \beta_2\mu_2(z_1) + \cdots + \alpha_n\lambda_n(z_1) - \beta_n\mu_n(z_1)$$
$$\hat{\phi}(z_2) = \alpha_1\lambda_1(z_2) - \beta_1\mu_1(z_2) + \alpha_2\lambda_2(z_2) - \beta_2\mu_2(z_2) + \cdots + \alpha_n\lambda_n(z_2) - \beta_n\mu_n(z_2)$$
$$\vdots$$
$$\hat{\phi}(z_{2n}) = \alpha_1\lambda_1(z_{2n}) - \beta_1\mu_1(z_{2n}) + \alpha_2\lambda_2(z_{2n}) - \beta_2\mu_2(z_{2n}) + \cdots + \alpha_n\lambda_n(z_{2n}) - \beta_n\mu_n(z_{2n})$$

This yields the following matrix equation.

$$\begin{bmatrix} \hat{\phi}(z_1) \\ \hat{\phi}(z_2) \\ \vdots \\ \hat{\phi}(z_{2n}) \end{bmatrix} = \begin{bmatrix} \lambda_1(z_1) & -\mu_1(z_1) & \lambda_2(z_1) & -\mu_2(z_1) & \cdots & \lambda_n(z_1) & -\mu_n(z_1) \\ \lambda_1(z_2) & -\mu_1(z_2) & \lambda_2(z_2) & -\mu_2(z_2) & \cdots & \lambda_n(z_2) & -\mu_n(z_2) \\ & & & & \vdots & & \\ \lambda_1(z_{2n}) & -\mu_1(z_{2n}) & \lambda_2(z_{2n}) & -\mu_2(z_{2n}) & \cdots & \lambda_n(z_{2n}) & -\mu_n(z_{2n}) \end{bmatrix} \begin{bmatrix} \alpha_1 \\ \beta_1 \\ \alpha_2 \\ \beta_2 \\ \vdots \\ \alpha_n \\ \beta_n \end{bmatrix}$$

4.5 Computational error estimation

In this section:

1. Error estimation using maximum modulus principle.
2. Measuring error using various norms.
3. Description of how the harmonic properties of error function informs the creation of more accurate CVBEM models.

A convenience of modeling with the CVBEM is that complex variable theory provides a simple technique for determining the maximum error of the CVBEM approximation function. Let ψ denote the exact solution of a given BVP (the following argument also applies when ϕ is the exact solution). Since the exact solution of the problem, given by ψ, is harmonic in Ω, and the imaginary part of the CVBEM approximation function, given by $\hat{\psi} = \text{Im}[\hat{\omega}]$, is harmonic in Ω, it follows that the error function, which is computed by the difference of these two harmonic functions, $\psi - \hat{\psi}$, is itself harmonic in Ω.

Consequently, by the maximum modulus principle for harmonic functions, $|\psi - \hat{\psi}|$ attains its maximum on $\partial\Omega$. Hence, the maximum error of the CVBEM approximation function occurs on $\partial\Omega$. Therefore, the magnitude of the error of the CVBEM approximation function within the problem domain in the ℓ_∞ norm is bounded by the maximum value of $|\psi - \hat{\psi}|$ on $\partial\Omega$.

Numerically, a reasonable approximation of $\max\limits_{\partial\Omega} |\psi - \hat{\psi}|$ can be obtained by computing the value of $|\psi - \hat{\psi}|$ at many locations along the problem boundary. As $|\psi - \hat{\psi}|$ is computed at more locations along the boundary, the estimation of $\max\limits_{\partial\Omega} |\psi - \hat{\psi}|$ will improve provided that the error evaluation points are reasonably spaced. For the results obtained in the subsequent sections, the error function was computed at 20,000 evenly spaced points on the problem boundary.

The four families are applied to the identical ideal fluid flow problem setting using collocation to determine the necessary complex constant coefficient values. In all cases tested, nodes and collocation points are positioned identically. To compare modeling error, a new computational measure is applied defined by $M = E/N$, where M is the computational measure value, E is the maximum computational departure between approximation and boundary condition value along the entire problem boundary (in absolute value), and N is the number of degrees of freedom used in the approximation (equivalently, the number of real-valued coefficients determined in the CVBEM modeling process for use in the approximation function linear combination.). Flow nets are developed for graphical comparison purposes. Computational results indicate that choice of basis function family primarily impacts the rapidity of convergence of the approximation function.

The error of the CVBEM approximation function can be determined by considering the difference between the CVBEM approximation function and the analytic potential solution (Equation (4.82) below). The difference between these two functions is the "error" function, and since it is the difference of two analytic functions over the problem domain and approximate boundary, it is itself analytic over that region. Thus, the error function attains its maximum (and minimum) values on the problem boundary by the Maximum Modulus Theorem. Therefore, in the case of a Dirichlet boundary value problem, modeling error in the approximation potential function can be assessed by examination of the maximum departure between the boundary values of the CVBEM approximation potential function and the boundary values of the analytic potential solution. It is noted that modeling error can be reduced by inserting additional collocation points in the vicinity of areas of large modeling error values.

From the Maximum Modulus Theorem, the maximum error of the CVBEM approximation function occurs on the boundary. Therefore, in order to determine the maximum error, the CVBEM approximation potential function was evaluated at 600 points along the boundary and the value of the CVBEM approximation potential function at each point was compared to the analytic potential solution given by $f(x, y) = x^2 - y^2$. The maximum error corresponded to the maximum departure (in absolute value) of the CVBEM approximation of the potential function from the analytic potential solution. The 600 test points were spaced evenly along the boundary with 200 test points located along the long ends of the rectangular boundary and 100 test points located along each of the short ends of the rectangular boundary.

Suppose $\omega(z) = \phi(x, y) + i\psi(x, y)$ is analytic within a given domain Ω. In Section 2.3, we showed that the real and imaginary parts of ω are harmonic wherever ω is analytic:

$$\nabla^2\phi = 0 \text{ and } \nabla^2\psi = 0 \quad \text{in } \Omega. \tag{4.92}$$

The original CVBEM basis functions were derived by using polynomial interpolation of given potential and streamline values along the problem boundary to generate an approximate boundary. If linear trial functions are used in the interpolation, then the CVBEM basis functions that result from numerically solving the Cauchy integral are products of complex linear polynomials with complex logarithmic functions. Likewise, when higher-order polynomial trial functions are used in the interpolation of the problem boundary, the resulting CVBEM approximation function is similarly composed of basis functions that are products of complex polynomials with complex logarithm functions. However, while the use of higher-order polynomial trial functions in the interpolation of the problem boundary is possible, in this research, the basis functions resulting from linear interpolation along the problem boundary are used.

It is noted that the term CVBEM refers to the general use of complex basis functions to solve boundary value problems, provided that the basis functions are analytic within the problem domain. Further, it is possible to mix families of basis functions to better fit particular problem geometries, although, such hybridizations are not considered in this research.

This research examines the solution success of families of basis functions in addition to those derived from the numerical solution of the Cauchy integral. All of the CVBEM approximation functions considered have the desirable property that they are well-defined and analytic throughout the problem domain as well as along the approximate boundary that is being modeled (see Johnson, et al. [27]). A CVBEM approximation function is a linear combination of members of the basis function family being used in the approximation. Since the basis functions examined in this research are analytic in the problem domain and along the boundary, the CVBEM approximation function is itself analytic in that region.

Chapter 5

CVBEM implementation in *Excel*

Bryce D. Wilkins

Contents

In this chapter, we develop two CVBEM models using *Excel*. In Section 5.2, the CVBEM procedure is used to model potential flow in a 90-degree bend. Then, in Section 5.3, the procedure is used to model potential flow over a half-cylinder.

Potential flow problems such as the two that will be examined in this chapter can readily be modeled using existing software packages for solving BVPs. While these packages are easy to use, their user-friendliness often obscures the underlying mechanics of the numerical methods being employed to solve the BVP. Therefore, the use of such packages oftentimes diminishes the learning experience that would otherwise be obtained by more direct handling of the computational underpinnings involved with methods such as the CVBEM. This realization is one of the primary motivations for developing the CVBEM models using *Excel*, which allows for the computational steps to be more explicitly displayed.

The goal of this chapter is to make the CVBEM methodology more transparent to the reader by providing two detailed implementations of the procedure. Accompanying *Excel* files are provided online for the reader's reference throughout this chapter.

5.1 Implementation

The CVBEM *Excel* program is implemented using five tabs in *Excel*. The tabs are as follows:

1. The **Problem Data** tab.
2. The **Coefficient Matrix** tab.
3. The **Approximation Function** tab.
4. The **Approximation** tab.
5. The **Plots** tab.

5.1.1 Problem Data tab

The **Problem Data** tab contains information related to the geometry of the problem domain. The user must specify the following data in this tab:

1. The length and height of the problem domain.
2. The coordinates of the computational nodes.

3. The coordinates of the collocation points.
4. The boundary conditions.

For the CVBEM models developed in this chapter, we will use collocation to determine the coefficients of the approximation functions. Therefore, for a CVBEM approximation function using n basis functions, it is necessary to specify the locations of $2n$ collocation points. This setup will lead to a matrix equation of size $2n \times 2n$ as specified in Section 4.4.

5.1.2 Coefficient Matrix tab

Next, the **Coefficient Matrix** tab sets up the matrix equation that will be used to determine the coefficients of the CVBEM approximation function. Recall, the CVBEM approximation function is a linear combination of analytic complex variable basis functions of the form:

$$\hat{\omega}(z) = \sum_{j=1}^{n} c_j g_j(z), \quad \text{where } c_j \in \mathbb{C}, \text{ and } g_j(z) \text{ is analytic.} \tag{5.1}$$

The purpose of the **Coefficient Matrix** tab is to compute the unknown coefficients $c_j \in \mathbb{C}$. Since these coefficients are complex numbers, each coefficient has both a real and an imaginary part: $c_j = \alpha_j + i\beta_j$. Therefore, there are two degrees of freedom that must be determined for each c_j; namely $\alpha_j \in \mathbb{R}$ and $\beta_j \in \mathbb{R}$. Additionally, the basis functions, $g_j(z)$, are complex variable functions and hence also have both a real and an imaginary part: $g_j(z) = \lambda_j(x,y) + i\mu_j(x,y)$. Thus, Equation (5.1) is equivalent to:

$$\begin{aligned}
\hat{\omega}(z) &= \sum_{j=1}^{n} c_j g_j(z) \\
&= \sum_{j=1}^{n} \left(\alpha_j + i\beta_j\right)\left(\lambda_j(x,y) + i\mu_j(x,y)\right) \\
&= \sum_{j=1}^{n} \left[\alpha_j\lambda_j(x,y) - \beta_j\mu_j(x,y) + i\alpha_j\mu_j(x,y) + i\beta_j\lambda_j(x,y)\right].
\end{aligned} \tag{5.2}$$

The real and imaginary parts of Equation (5.2) are resolved as follows:

$$\begin{aligned}
\hat{\phi}(x,y) &= \text{Re}\left[\hat{\omega}(z)\right] = \sum_{j=1}^{n} \alpha_j\lambda_j(x,y) - \beta_j\mu_j(x,y) \\
\hat{\psi}(x,y) &= \text{Im}\left[\hat{\omega}(z)\right] = \sum_{j=1}^{n} \alpha_j\mu_j(x,y) + \beta_j\lambda_j(x,y).
\end{aligned} \tag{5.3}$$

Note 5.1.2.1

Within the context of potential flow, $\hat{\phi}$ and $\hat{\psi}$ are interpreted physically as the CVBEM approximation of the potential and stream functions, respectively. In two spatial dimensions, the functions $\hat{\phi}$ and $\hat{\psi}$ are conjugate harmonic functions. Consequently, their level curves are orthogonal. The orthogonality of their level curves is what allows for the production of the flow nets that are commonly used to visualize potential flow.

In Equation (5.3), the unknown quantities of $\hat{\phi}$ and $\hat{\psi}$ are the real numbers α_j and β_j, $j = 1, \ldots, n$. That is, $\hat{\phi}$ and $\hat{\psi}$ have the same coefficients. As a result, it is only necessary to determine the coefficients for one of the two functions. Then, the coefficients that are computed in the determination of one of the two functions can be used to evaluate the other function.

Note 5.1.2.2

Since $\hat{\phi}$ and $\hat{\psi}$ have the same coefficients, it is not necessary to do any type of post-processing to determine the corresponding conjugate stream function after the potential function has been approximated using the CVBEM.

Now, we will address the process of determining the α_j and β_j. The problems examined in the following sections are formulated as Dirichlet boundary value problems. These are problems in which the values of the target potential function are known along the problem boundary. Together, the α_j and β_j constitute $2n$ degrees of freedom. In this work, we use collocation to fit the CVBEM approximation function to the boundary conditions, which means it is necessary to specify the locations of $2n$ points along the problem boundary at which the boundary conditions will be enforced. These points are denoted $(x_k, y_k) \in \partial\Omega$, and the corresponding boundary conditions at these points are denoted $\phi(x_k, y_k)$, $k = 1, \ldots, 2n$. This leads to a system of $2n$ equations in $2n$ unknowns where the kth equation is given by

$$\hat{\phi}(x_k, y_k) = \sum_{j=1}^{n} \alpha_j \lambda_j(x_k, y_k) - \beta_j \mu_j(x_k, y_k) = \phi(x_k, y_k). \tag{5.4}$$

Equivalently,

$$\hat{\phi}(x_k, y_k) = \underbrace{\sum_{j=1}^{n} \alpha_j \lambda_j(x_k, y_k) - \sum_{j=1}^{n} \beta_j \mu_j(x_k, y_k)}_{\text{CVBEM approximation function}} = \underbrace{\phi(x_k, y_k)}_{\substack{\text{Known}\\ \text{boundary}\\ \text{condition}}} . \tag{5.5}$$

Note 5.1.2.3

Equations (5.4) and (5.5) are obtained by evaluating $\hat{\phi}(x, y)$ from Equation (5.3) at each of the $2n$ collocation points and setting the result equal to the boundary condition at that point.

Therefore, the system of equations that needs to be solved is as follows.

$$\begin{aligned}
\phi(x_1, y_1) &= \alpha_1 \lambda_1(x_1, y_1) + \alpha_2 \lambda_2(x_1, y_1) + \cdots + \alpha_n \lambda_n(x_1, y_1) \\
&\quad - \beta_1 \mu_1(x_1, y_1) - \beta_2 \mu_2(x_1, y_1) - \cdots - \beta_n \mu_n(x_1, y_1) \\
\phi(x_2, y_2) &= \alpha_1 \lambda_1(x_2, y_2) + \alpha_2 \lambda_2(x_2, y_2) + \cdots + \alpha_n \lambda_n(x_2, y_2) \\
&\quad - \beta_1 \mu_1(x_2, y_2) - \beta_2 \mu_2(x_2, y_2) - \cdots - \beta_n \mu_n(x_2, y_2) \\
&\ \vdots \\
\phi(x_{2n}, y_{2n}) &= \alpha_1 \lambda_1(x_{2n}, y_{2n}) + \alpha_2 \lambda_2(x_{2n}, y_{2n}) + \cdots + \alpha_n \lambda_n(x_{2n}, y_{2n}) \\
&\quad - \beta_1 \mu_1(x_{2n}, y_{2n}) - \beta_2 \mu_2(x_{2n}, y_{2n}) - \cdots - \beta_n \mu_n(x_{2n}, y_{2n})
\end{aligned} \tag{5.6}$$

This yields the following matrix equation.

$$
\underbrace{\begin{bmatrix}
\lambda_1(x_1,y_1) & \lambda_2(x_1,y_1) & \cdots & \lambda_n(x_1,y_1) & -\mu_1(x_1,y_1) & -\mu_2(x_1,y_1) & \cdots & -\mu_n(x_1,y_1) \\
\lambda_1(x_2,y_2) & \lambda_2(x_2,y_2) & \cdots & \lambda_n(x_2,y_2) & -\mu_1(x_2,y_2) & -\mu_2(x_2,y_2) & \cdots & -\mu_n(x_2,y_2) \\
\lambda_1(x_3,y_3) & \lambda_2(x_3,y_3) & \cdots & \lambda_n(x_3,y_3) & -\mu_1(x_3,y_3) & -\mu_2(x_3,y_3) & \cdots & -\mu_n(x_3,y_3) \\
\lambda_1(x_4,y_4) & \lambda_2(x_4,y_4) & \cdots & \lambda_n(x_4,y_4) & -\mu_1(x_4,y_4) & -\mu_2(x_4,y_4) & \cdots & -\mu_n(x_4,y_4) \\
 & & & \vdots & & & & \\
\lambda_1(x_{2n},y_{2n}) & \lambda_2(x_{2n},y_{2n}) & \cdots & \lambda_n(x_{2n},y_{2n}) & -\mu_1(x_{2n},y_{2n}) & -\mu_2(x_{2n},y_{2n}) & \cdots & -\mu_n(x_{2n},y_{2n}).
\end{bmatrix}}_{\text{The basis functions evaluated at the location of each of the collocation points}}
\underbrace{\begin{bmatrix}
\alpha_1 \\ \alpha_2 \\ \vdots \\ \alpha_n \\ \beta_1 \\ \beta_2 \\ \vdots \\ \beta_n
\end{bmatrix}}_{\substack{\text{Unknown} \\ \text{coefficients} \\ \text{to be} \\ \text{determined}}}
=
\underbrace{\begin{bmatrix}
\phi(x_1,y_1) \\ \phi(x_2,y_2) \\ \phi(x_3,y_3) \\ \phi(x_4,y_4) \\ \vdots \\ \phi(x_{2n},y_{2n})
\end{bmatrix}}_{\substack{\text{Dirichlet} \\ \text{boundary} \\ \text{conditions}}}
\tag{5.7}
$$

The use of collocation when determining the $2n$ degrees of freedom ensures that the CVBEM approximation function will be equal to the exact solution at the $2n$ locations on the problem boundary where the boundary conditions are specified. Thus, within the context of the approximate boundary, using collocation ensures that there will be at least $2n$ locations at which the approximate boundary intersects the true boundary; namely, at the collocation points.

While collocation is used in this work, it is also possible to use other techniques, such as least squares, for determining the coefficients. Least squares would be useful in a situation in which the user desires to use the boundary data at more than $2n$ locations when determining the $2n$ degrees of freedom.

Note 5.1.2.4

The matrix corresponding to the equations given in Equation (5.4) is dense as opposed to the sparse matrices frequently encountered in finite difference and finite element analyses. The reason for this is that the CVBEM basis functions have global support, whereas the basis functions commonly used in the finite difference and finite element methods have local support.

Note 5.1.2.5

For information on the handling of mixed boundary value problems, see [28].

There are seven major features of the **Coefficient Matrix** tab. These features are numbered and discussed in the list below.

1. $A \in \mathbb{R}^{2n \times 2n}$: the coefficient matrix. The values of the coefficient matrix correspond to evaluating the real and imaginary parts of the basis functions at each of the collocation points. In particular, each column of A corresponds to a particular basis function (either the real or imaginary part of the basis function), and each row of A corresponds to a particular collocation point.

   Recall the definition of the complex coefficients of the CVBEM approximation function: $c_j = \alpha_j + i\beta_j$. The coefficient matrix is constructed so that the first n columns of A correspond to equations for α_j, $j = 1, \ldots, n$. And columns $n + 1$ to $2n$ of A correspond to equations for β_j, $j = 1, \ldots, n$.

The basis functions used in these *Excel* models are of the "distributed sources" type introduced in Equation (4.79), where each point $z_j \in \mathbb{C}$ represents one of the distributed sources. Therefore, the CVBEM approximation function is of the form:

$$\hat{\omega}(z) = \sum_{j=1}^{n} c_j \frac{1}{z - z_j}. \tag{5.8}$$

Consequently, the entries of the coefficient matrix, A_{kj}, are computed as follows:

$$A_{kj} = \begin{cases} \text{Re}\left(\frac{1}{z_k - z_j}\right), & 1 \le j \le n \\ \\ \text{Im}\left(\frac{1}{z_k - z_j}\right), & n + 1 \le j \le 2n, \end{cases} \tag{5.9}$$

where $z_k = x_k + iy_k$, $k = 1, \ldots, 2n$, denotes the index of the kth collocation point, (x_k, y_k).

2. $x \in \mathbb{R}^{2n}$: the vector of unknown coefficients. These are the values α_j and β_j that will form the complex coefficients of the CVBEM approximation function.

Note 5.1.2.6

The main computational effort of the *Excel* CVBEM program is in determining the unknown coefficients of the CVBEM approximation function.

Given the construction of A in Equation (5.9), in which the first n columns correspond to the real parts of the basis functions and columns $n + 1$ to $2n$ correspond to the imaginary parts of the basis functions, the coefficient vector is structured as follows:

$$x = \begin{bmatrix} \alpha_1 & \alpha_2 & \cdots & \alpha_n & \beta_1 & \beta_2 & \cdots & \beta_n \end{bmatrix}^{\top} = \begin{bmatrix} \alpha & \beta \end{bmatrix}^{\top} \tag{5.10}$$

3. $Ax = \hat{b}$: the CVBEM approximation of the boundary conditions. This approximation depends on the values of the coefficients in the $x = \begin{bmatrix} \alpha & \beta \end{bmatrix}^{\top}$ vector. This column is computed using the MMULT() command to multiply A and x. The result, $\hat{b} \in \mathbb{R}^{2n}$, is the CVBEM approximation of the boundary conditions at each of the collocation points.

4. $b \in \mathbb{R}^{2n}$: the target boundary values. This data is specified in the "Boundary Condition" column of the **Problem Data** tab.

5. $\varepsilon = \left|\hat{b} - b\right| \in \mathbb{R}^{2n}$: the error of the CVBEM approximation function. The absolute error of the CVBEM approximation function is computed at each of the collocation points.

6. $\varepsilon^2 \in \mathbb{R}^{2n}$: the square of the approximation error. The square of the absolute error of the CVBEM approximation function is evaluated at each of the collocation points. This is a vector where the kth element is given by $\left(\varepsilon^2\right)_k = \left(\varepsilon_k\right)^2$.

7. $\text{SSE} \in \mathbb{R}$: the sum of squared errors. This is equivalent to $\varepsilon^{\top}\varepsilon$. The sum of squared errors is used as the objective function when determining the coefficient vector. The goal is to find the coefficients that minimize this quantity.

   Two approaches to determine the coefficients of the CVBEM approximation function are as follows:

   (a) The built-in MINVERSE() function can be used to compute A^{-1}. Then, the coefficients are computed as $x = A^{-1}b$. The difficulty with this approach is that the numerical computation of A^{-1} may not be precise enough to accurately determine the values of x.

   (b) Alternatively, the *Excel* "Solver" add-in can be used to determine the coefficients of the CVBEM approximation function. This determines x as the solution to an optimization problem. The objective function is the sum of squared errors, $\varepsilon^{\top}\varepsilon$. The variable cells are the cells constituting the coefficient vector, x. This is the approach we recommend, and the Solver window is shown in Figure 5.1.

5.1.3 Approximation function tab

After the coefficients are computed in the **Coefficient Matrix** tab, the **Approximation Function** tab is used to evaluate the CVBEM approximation function at various sampling points in the problem domain. For a given point $z_0 = x_0 + iy_0 \in \Omega$, the CVBEM approximation of the potential function is

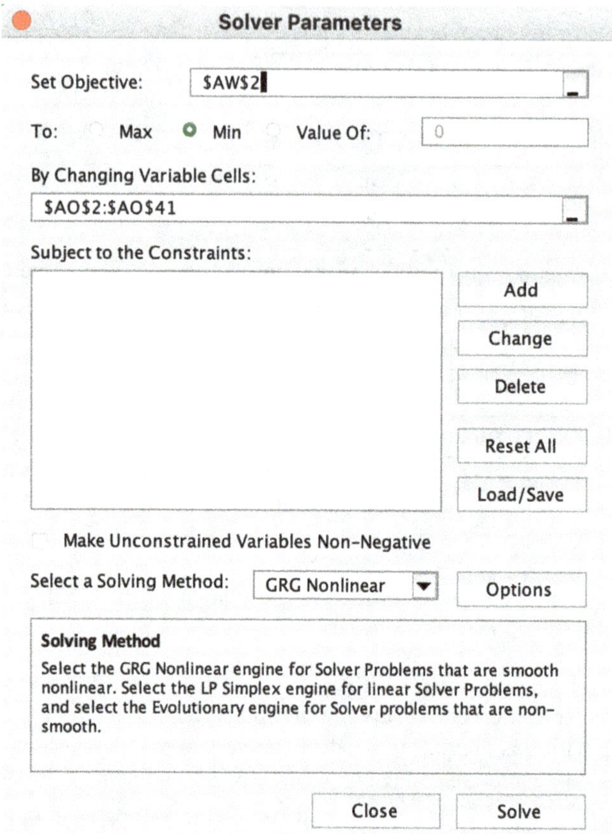

Figure 5.1: Using the Microsoft ®*Excel* Solver to determine the coefficients of the CVBEM approximation function. The objective function refers to the cell in which the sum of squared errors is calculated. The variable cells are the cells constituting the coefficient vector, x. This is a minimization problem because the goal is to determine the coefficients of the CVBEM approximation function that minimizes the Euclidean distance between the values of the CVBEM approximation function and the target boundary conditions. In this case, the sum of squared errors cell is $AW$2, and the cells of the coefficients vector are contained in $AO$2:$AO$41. Used with permission from Microsoft.

computed as:

$$\hat{\phi}(x_0, y_0) = \sum_{j=1}^{n} \alpha_j \mathrm{Re}\left(\frac{1}{z_0 - z_j}\right) - \beta_j \mathrm{Im}\left(\frac{1}{z_0 - z_j}\right), \quad z_0 \in \Omega. \quad (5.11)$$

The CVBEM approximation of the stream function at z_0 is computed as:

$$\hat{\psi}(x_0, y_0) = \sum_{j=1}^{n} \alpha_j \mathrm{Im}\left(\frac{1}{z_0 - z_j}\right) + \beta_j \mathrm{Re}\left(\frac{1}{z_0 - z_j}\right), \quad z_0 \in \Omega. \tag{5.12}$$

Note 5.1.3.1

The equations given in (5.11) and (5.12) are equivalent to the equations given in (5.3).

Note 5.1.3.2

Since the CVBEM approximation function is an analytic function everywhere except at the source points, z_j, the approximation function can be evaluated in the exterior of the problem domain. While this may be useful for some applications, it is not necessary for the examples discussed in the following sections. Specifically, the following examples are most interested in developing flow nets in the interior of the problem domain. For this reason, we only consider evaluating the CVBEM approximation function in the interior of the problem domain.

5.1.4 Approximation tab

The **Approximation** tab is used to reshape the data from the **Approximation Function** tab in order to develop contours of the real and imaginary parts of the CVBEM approximation function. The **Approximation** tab consists of two matrices:

1. The first matrix consists of values obtained by evaluating the real part of the CVBEM approximation function at various points in the problem domain. These values are used to produce the contour plot of potential values in the **Plots** tab.

2. The second matrix consists of values obtained by evaluating the imaginary part of the CVBEM approximation function at the same points in the problem domain. These values are used to produce the contour plot of streamline values in the **Plots** tab.

5.1.5 Plots tab

The **Plots** tab contains a depiction of the problem geometry, including the locations of computational nodes and collocation points. This tab also includes a flow net representing the CVBEM approximation of the potential function as well as a flow net representing the CVBEM approximation of the stream function.

The following instructions may be used to produce contour plots in *Excel*.

1. Select the data to be plotted. The data needs to be in a rectangular array.
2. Insert → Chart → Surface.
3. Delete the legend.
4. Delete the axis scales except for the scale named "Vertical (Value) Axis."
5. Select Add Chart Element → Gridlines → Deselect "Primary Major Horizontal."
6. Right click on the graph → 3-D Rotation → Change the following values:
   - X Rotation = 270
   - Y Rotation = 90
   - Perspective = 0.1
   - Note: to change the aspect ratio, deselect Autoscale and change the following values: Depth (% of base) and Height (% of base)
7. Compute the minimum and maximum values of the data set to be plotted. You can do this using the *Excel* commands: MIN( ) and MAX( ).
8. Right click on the "Vertical (Value) Axis" → Bounds → Change Minimum and Maximum bounds to match the minimum and maximum values of the data set.
9. Right click on the "Vertical (Value) Axis" → Bounds → Change Major and Minor units to increase the number of contour lines.
10. Delete "Vertical (Value) Axis" scale.

5.2 Potential flow in a 90-degree bend

In this section, we consider the standard benchmark problem of potential flow in a 90-degree bend. CVBEM results from modeling this problem using the programs MATLAB and MATHEMATICA were shown in Section 4.3.5.

5.2.1 The Problem Data tab

In this example, the problem domain is taken to be $\Omega = [0,1] \times [0,1]$. Given this definition, we specify the Length parameter as 1 and the Height parameter as 1, as indicated in Table 5.1.

Furthermore, we are going to create a CVBEM model with $n = 20$ terms in the approximation function. Therefore, this requires the specification of $n = 20$ computational nodes and $2n = 40$ collocation points.

We want the computational nodes to be uniformly spaced in a circle around the problem domain. A circle consists of 2π radians, and we desire that the circle is evenly discretized by 20 points (these are the computational nodes). Therefore, the distance between successive nodes on the circle is given by $2\pi/20$. This quantity is referred to as dtheta in Table 5.1.

Likewise, we want the collocation points to be uniformly spaced along the perimeter of the problem boundary. The perimeter is 4 units long, and we desire that the perimeter is evenly discretized by 40 points (these are the collocation points). Therefore, the distance between successive collocation points on the problem boundary is given by $4/40 = 0.1$. Since the problem geometry is a square, this quantity is used as both dx and dy in Table 5.1.

Excel Implementation 5.2.1.1

Length	1
Height	1
dx (for collocation points)	0.1
dy (for collocation points)	0.1
dtheta (for nodes)	=2*PI()/20

Table 5.1: The user will input problem data into a table of this form in the **Problem Data** tab. The dtheta parameter is used to specify 20 computational nodes in a circular pattern in the exterior of the problem domain. The dx and dy parameters are used to evenly space 40 collocation points around the problem boundary.

This setup is depicted in Figure 5.2 and can be found in the **Plots** tab of the accompanying *Excel* workbook.

The "Boundary Condition" column of the **Problem Data** tab is found by evaluating the target potential function at each of the collocation points. For this BVP, the boundary conditions are given as:

$$\phi(x,y) = \mathrm{Re}\left(z^2\right) = x^2 - y^2, \quad (x,y) \in \partial\Omega. \tag{5.13}$$

Therefore, the boundary conditions are computed as $x_k^2 - y_k^2$, where $(x_k, y_k) \in \partial\Omega$ denotes the location of the kth boundary condition.

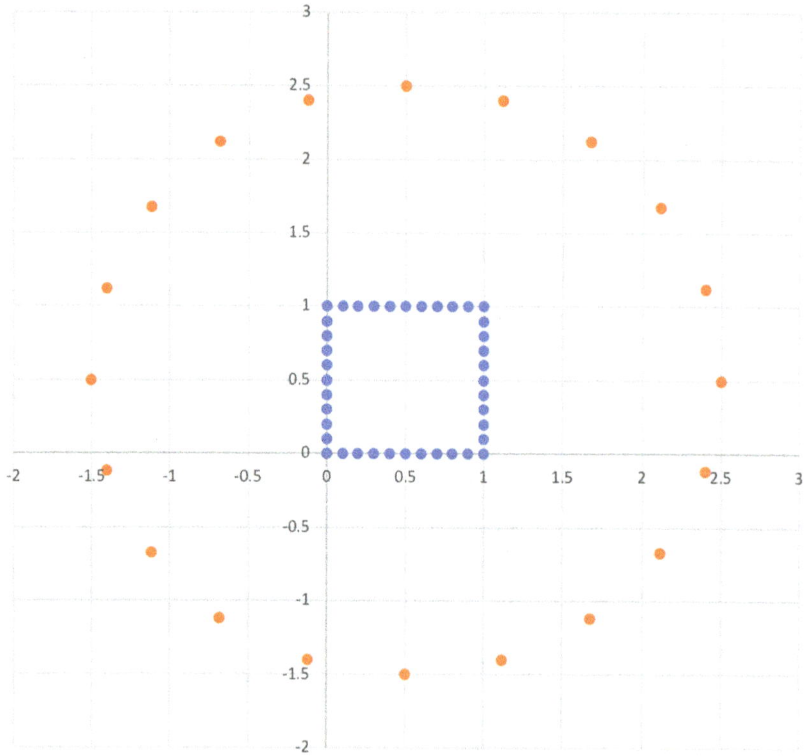

Figure 5.2: Locations of nodes and collocation points used in the CVBEM model. Nodes are depicted as orange points, and collocation points are depicted as blue points. The physical phenomena being modeled is potential flow in a 90-degree bend.

Excel Implementation 5.2.1.2

Figures 5.4 and 5.5 show the complex coordinates for the computational nodes used in the CVBEM approximation function for this problem. Since the basis functions used in this model are of the form $\frac{1}{z-z_j}$, the computational nodes are identified as the poles of the basis functions.

We will give the general equation for each row shown in these figures. Define the following:

$$\begin{aligned} \theta_k &\equiv \text{the } k^{\text{th}} \text{ cell of the } \Theta \text{ row} \\ x_k &\equiv \text{the } k^{\text{th}} \text{ cell of the } x\text{-coordinate row} \\ y_k &\equiv \text{the } k^{\text{th}} \text{ cell of the } y\text{-coordinate row} \\ z_k &\equiv \text{the } k^{\text{th}} \text{ cell of the complex coordinate row} \end{aligned} \tag{5.14}$$

Given these variable names, the general equations for each of the rows shown in Figures 5.4 and 5.5 are as follows:

1. Θ: $\theta_{i+1} = \theta_i +$ dtheta, where $\theta_1 =$ dtheta, as defined in Table 5.1.
2. x-coordinate of poles: $x_i = 2\text{COS}(\theta_i) + 0.5$, where COS() is the built-in cosine function.
3. y-coordinate of poles: $y_i = 2\text{SIN}(\theta_i) + 0.5$, where SIN() is the built-in sine function.
4. Complex coordinate of poles: $z_i = \text{COMPLEX}(x_i, y_i)$.

Note: the scalar of 2 appearing before COS(θ_i) and SIN(θ_i) establishes the fact that the radius for the circle of nodes is 2. Furthermore, the shift of 0.5 in these equations means that the circle of nodes is centered at $(0.5, 0.5)$, which is the center of the problem domain.

Excel Implementation 5.2.1.3

The table shown in Figure 5.3 contains the complex coordinates of the collocation points as well as the value of the boundary condition at each of the collocation points. The first collocation point is located at the origin, $(0, 0)$. Then, the collocation points are placed counterclockwise along the problem domain, $\Omega = [0,1] \times [0,1]$. The spacings of the collocation points, dx and dy, correspond to the values given in Table 5.1. Define the following:

$$\begin{aligned} x_k &\equiv \text{the } k^{\text{th}} \text{ cell of the } x\text{-coordinate column} \\ y_k &\equiv \text{the } k^{\text{th}} \text{ cell of the } y\text{-coordinate column} \\ z_k &\equiv \text{the } k^{\text{th}} \text{ cell of the complex coordinate column} \\ \phi_k &\equiv \text{the } k^{\text{th}} \text{ cell of the boundary condition column} \end{aligned} \tag{5.15}$$

Given these variable names, the general equations for the complex coordinate column and the boundary condition column shown in Figure 5.3 is as follows:

1. Complex coordinate column: $z_k = \text{COMPLEX}(x_k, y_k)$
2. Boundary condition column: $\phi_k = \text{IMREAL}(\text{IMPOWER}(z_k, 2))$

The equation for the boundary condition column is defined by the boundary conditions of the problem. In this case, the boundary conditions are $\phi(x,y) = \text{Re}\left(z^2\right)$, $(x,y) \in \partial\Omega$.

x-coordinate of collocation points	y-coordinate of collocation points	Complex coordinate of collocation points	Boundary Condition
0	0	0	0
0.1	0	0.1	0.01
0.2	0	0.2	0.04
0.3	0	0.3	0.09
0.4	0	0.4	0.16
0.5	0	0.5	0.25
0.6	0	0.6	0.36
0.7	0	0.7	0.49
0.8	0	0.8	0.64
0.9	0	0.9	0.81
1	0	1	1
1	0.1	1+0.1i	0.99
1	0.2	1+0.2i	0.96
1	0.3	1+0.3i	0.91
1	0.4	1+0.4i	0.84
1	0.5	1+0.5i	0.75
1	0.6	1+0.6i	0.64
1	0.7	1+0.7i	0.51
1	0.8	1+0.8i	0.36
1	0.9	1+0.9i	0.19
1	1	1+1i	1E-16
0.9	1	0.9+i	-0.19
0.8	1	0.8+i	-0.36
0.7	1	0.7+i	-0.51
0.6	1	0.6+i	-0.64
0.5	1	0.5+i	-0.75
0.4	1	0.4+i	-0.84
0.3	1	0.3+i	-0.91
0.2	1	0.2+i	-0.96
0.1	1	0.1+i	-0.99
0	1	i	-1
0	0.9	0.9i	-0.81
0	0.8	0.8i	-0.64
0	0.7	0.7i	-0.49
0	0.6	0.6i	-0.36
0	0.5	0.5i	-0.25
0	0.4	0.4i	-0.16
0	0.3	0.3i	-0.09
0	0.2	0.2i	-0.04
0	0.1	0.1i	-0.01

Figure 5.3: Table in the *Excel* program specifying the x- and y-coordinates of the collocation points. In the third column, the real coordinates of the collocation points are converted to the corresponding point in the complex plane. In the final column, we specify the value of the boundary condition at the particular collocation point. For this problem, the boundary conditions are $\phi_k(x_k, y_k) = \text{Re}\left(z_k^2\right)$, $(x_k, y_k) \in \partial\Omega$.

θ	0.3142	0.6283	0.9425	1.2566	1.5708	1.885	2.1991	2.5133	2.8274	3.1416
x-coordinate of poles:	2.4021	2.118	1.6756	1.118	0.5	-0.118	-0.6756	-1.118	-1.4021	-1.5
y-coordinate of poles:	1.118	1.6756	2.118	2.4021	2.5	2.4021	2.118	1.6756	1.118	0.5
Complex coordinate of poles:	2.40211:	2.11803:	1.67557(	1.11803:	0.5+2.5i	-0.11803	-0.67557	-1.11803	-1.40211	-1.5+0.5

Figure 5.4: Table in *Excel* specifying the x- and y-coordinates of the computational nodes. The first row is a parameter named θ, which is used to orient the computational nodes in a circle, as illustrated in Figure 5.2. In the fourth row, the real coordinates of the computational nodes are converted to the corresponding point in the complex plane.

θ	3.4558	3.7699	4.0841	4.3982	4.7124	5.0265	5.3407	5.6549	5.969	6.2832
x-coordinate of poles:	-1.4021	-1.118	-0.6756	-0.118	0.5	1.118	1.6756	2.118	2.4021	2.5
y-coordinate of poles:	-0.118	-0.6756	-1.118	-1.4021	-1.5	-1.4021	-1.118	-0.6756	-0.118	0.5
Complex coordinate of poles:	-1.40211	-1.11803	-0.67557	-0.11803	0.5-1.5i	1.11803:	1.67557(	2.11803:	2.40211:	2.5+0.5i

Figure 5.5: Continuation of the data presented in Figure 5.4.

5.2.2 The Coefficient Matrix tab

The **Coefficient Matrix** tab consists of the following parts:

1. $\boxed{A \in \mathbb{R}^{40\times 40}:}$ this matrix contains the real and imaginary parts of each of the 20 basis functions evaluated at each of the 40 collocation points. It is best to think of this matrix in block form, as follows:

$$A = \begin{bmatrix} \underbrace{A_1}_{\substack{\text{The real}\\ \text{part of the}\\ \text{basis functions}}} & \underbrace{A_2}_{\substack{\text{The imaginary}\\ \text{part of the}\\ \text{basis functions}}} \end{bmatrix}, \quad A_1, A_2 \in \mathbb{R}^{40\times 20}. \tag{5.16}$$

Let (x_i, y_i) for $i = 1, \ldots, 40$ denote the locations of the collocation points. Then, $A_1(i,j) = \lambda_j(x_i, y_i)$ and $A_2(i,j) = -\mu_j(x_i, y_i)$ (these expressions are based on the equation for $\hat{\phi}(x,y)$ in Equation (5.3)).

$$\begin{aligned} A_1 &= \begin{bmatrix} \lambda_1(x_1,y_1) & \lambda_2(x_1,y_1) & \cdots & \lambda_{20}(x_1,y_1) \\ \lambda_1(x_2,y_2) & \lambda_2(x_2,y_2) & \cdots & \lambda_{20}(x_2,y_2) \\ \vdots & \vdots & \ddots & \vdots \\ \lambda_1(x_{40},y_{40}) & \lambda_2(x_{40},y_{40}) & \cdots & \lambda_{20}(x_{40},y_{40}) \end{bmatrix} \\ A_2 &= \begin{bmatrix} -\mu_1(x_1,y_1) & -\mu_2(x_1,y_1) & \cdots & -\mu_{20}(x_1,y_1) \\ -\mu_1(x_2,y_2) & -\mu_2(x_2,y_2) & \cdots & -\mu_{20}(x_2,y_2) \\ \vdots & \vdots & \ddots & \vdots \\ -\mu_1(x_{40},y_{40}) & -\mu_2(x_{40},y_{40}) & \cdots & -\mu_{20}(x_{40},y_{40}) \end{bmatrix} \end{aligned} \tag{5.17}$$

There are at least two ways to calculate $-\mu_j(x_i, y_i)$. One way is as $-\text{Im}\left(\frac{1}{z_i - z_j}\right)$, where $z_i = x_i + iy_i$ and $z_j = x_j + iy_j$ (note: the difference between the index i and the imaginary number should be clear from the context). The alternative way is as $\text{Re}\left(\frac{i}{z_i - z_j}\right)$. We use the latter option in the *Excel* implementation demonstrated in this work.

***Excel* Implementation 5.2.2.1**

A general cell of the matrix $A = \begin{bmatrix} A_1 & A_2 \end{bmatrix}$ is computed as follows:

$$\begin{aligned} A_1(i,j) &= \text{IMREAL(IMDIV(1,IMSUB(}z_i,z_j\text{)))} \\ A_2(i,j) &= \text{IMREAL(IMPRODUCT(COMPLEX(0,1),...} \\ &\quad \text{IMDIV(1,IMSUB(}z_i,z_j\text{))))} \end{aligned} \tag{5.18}$$

2. $\boxed{x \in \mathbb{R}^{20}:}$ this vector consists of the coefficients of the CVBEM approximation function. The user has to initialize this vector. A choice of all 1's is usually sufficient. Then, the built-in "Solver" application is used to determine the values of these coefficients by optimization. The "Solver" application can be configured as shown in Figure 5.1.

3. $\boxed{Ax = \hat{b} \in \mathbb{R}^{20}:}$ this vector contains the CVBEM approximation of the target boundary conditions.

***Excel* Implementation 5.2.2.2**

The CVBEM approximation of the target boundary conditions is computed using the MMULT() function. In this case, the A matrix is stored in the cells A2:AN41 and the x vector is stored in the cells AO2:AO41. Therefore, $\hat{b}$ is calculated as:

$$\hat{b} = \{\ \text{MMULT(A2:AN41,AO2:AO41)}\ \} \tag{5.19}$$

Note: the *Excel* expression for $\hat{b}$ is surrounded by curly braces. This syntax denotes that a matrix operation is being used. In order to perform a matrix operation like this, it is necessary to enter the command slightly differently. The following instructions can be used:

(a) Select the region where the results are to be stored. In this case, the results are to be stored in the column labeled $Ax = \hat{b}$. Therefore, we select the cells AP2:AP41.

(b) Enter the following: =MMULT(A2:AN41,AO2:AO41).

(c) Here is the important step. After entering the command, it is not enough to just press "Enter" on the keyboard. Instead, it is necessary to press "Control + Shift + Enter".

4. $\boxed{b \in \mathbb{R}^{20}:}$ this vector contains the target boundary conditions. The data in this vector is automatically copied from the "Boundary Condition" column in the **Problem Data** tab.

5. $\boxed{\text{Absolute error:}}$ this vector is the absolute difference between $\hat{b}$ and b, which is computed as $\varepsilon = \left|\hat{b} - b\right|$.

> ***Excel* Implementation 5.2.2.3**
>
> Define the following:
>
> $$\begin{aligned} \hat{b}_k &\equiv \text{the } k^{\text{th}} \text{ cell of the } Ax = \hat{b} \text{ column} \\ b_k &\equiv \text{the } k^{\text{th}} \text{ cell of the } b \text{ column} \\ \varepsilon_k &\equiv \text{the } k^{\text{th}} \text{ cell of the Absolute Error column} \end{aligned} \tag{5.20}$$
>
> Given these variable names, the general equation for the cells in the Absolute Error column is as follows:
>
> $$\varepsilon_k = \text{ABS}(\hat{b}_k - b_k) \tag{5.21}$$

6. $\boxed{\text{Error Squared:}}$ this vector contains the square of each element in ε.

> ***Excel* Implementation 5.2.2.4**
>
> Define the following:
>
> $$\begin{aligned} \varepsilon_k &\equiv \text{the } k^{\text{th}} \text{ cell of the Absolute Error column} \\ \varepsilon_k^2 &\equiv \text{the } k^{\text{th}} \text{ cell of the Error Squared column} \end{aligned} \tag{5.22}$$
>
> Given these variable names, the general equation for the cells in the Error Squared column is as follows:
>
> $$\varepsilon_k^2 = \text{POWER}(\varepsilon_k, 2) \tag{5.23}$$

7. $\boxed{\text{Sum of Squared Errors:}}$ this cell is equivalent to $\varepsilon^{\top}\varepsilon$ and to $\sum_k \varepsilon_k^2$. This cell is used as the objective cell for the "Solver" application.

***Excel* Implementation 5.2.2.5**

Let SSE denote the sum of squared errors. For this example problem, the ε^2 vector is stored in the cells AT2:AT41. Therefore, the Sum of Squared Errors cell is calculated as follows:

$$\text{SSE} = \text{SUM(AT2:AT41)} \tag{5.24}$$

5.2.3 The Approximation Function tab

The **Approximation Function** tab is used to evaluate the CVBEM approximation function at various points in the interior of the problem domain. These evaluated values are used to create the contour plots in the **Plots** tab. The features of this tab are as follows:

1. Evaluation point: z =. This is the point at which the CVBEM approximation function is to be evaluated. It is used as the "Column input cell" in *Excel*'s built-in "Data Table" program.

2. Coefficients. These are the coefficients that were obtained by optimization in the **Coefficient Matrix** tab.

3. REAL PART: Basis Functions at Evaluation Point. This is the real part of each of the basis functions evaluated at the evaluation point given in cell J2. Let $z_0 = x_0 + iy_0$ represent the entry in J2. Recall the equation for the real part of the CVBEM approximation function:

$$\hat{\phi}(x_0, y_0) = \sum_{j=1}^{n} \alpha_j \text{Re}\left(\frac{1}{z_0 - z_j}\right) - \beta_j \text{Im}\left(\frac{1}{z_0 - z_j}\right), \quad z_0 \in \Omega. \tag{5.25}$$

***Excel* Implementation 5.2.3.1**

Let z_j denote the location of the jth source point (i.e. the pole of the jth basis function), $1 \leq j \leq 20$. The first 20 cells in this row correspond to evaluating $\text{Re}\left(\frac{1}{z_0 - z_j}\right)$. These cells are computed as follows:

$$= \text{IMREAL(IMDIV(1,IMSUB(\$J\$2,}z_j\text{)))} \tag{5.26}$$

The next 20 cells in this row correspond to evaluating $-\text{Im}\left(\frac{1}{z_0-z_j}\right)$. These cells are computed as follows:

$$= \text{IMREAL(IMPRODUCT(COMPLEX(0,1),...}\\ \text{IMDIV(1,IMSUB(\$J\$2,}z_j\text{))))} \tag{5.27}$$

4. Real Product. Again, let z_0 denote the evaluation point given in cell J2. Recall that the real part of the CVBEM approximation function is of the form:

$$\hat{\phi}(x_0, y_0) = \sum_{j=1}^{n} \alpha_j \text{Re}\left(\frac{1}{z_0 - z_j}\right) - \beta_j \text{Im}\left(\frac{1}{z_0 - z_j}\right). \tag{5.28}$$

This row is used to compute each of the products given by $\alpha_j \text{Re}\left(\frac{1}{z_0-z_j}\right)$ and $-\beta_j \text{Im}\left(\frac{1}{z_0-z_j}\right)$ for each $1 \le j \le 20$.

***Excel* Implementation 5.2.3.2**

Define the following:

$$\begin{aligned} p_k &\equiv \text{the } k^{\text{th}} \text{ cell of the “Coefficients” row} \\ q_k &\equiv \text{the } k^{\text{th}} \text{ cell of the “REAL PART” row} \\ r_k &\equiv \text{the } k^{\text{th}} \text{ cell of the “Real Product” row} \end{aligned} \tag{5.29}$$

Given these variable names, the general equation for the “Real Product” row is as follows:

$$r_k = \text{PRODUCT}(p_k, q_k) \tag{5.30}$$

5. Potential Approximation. This is the CVBEM approximation of the potential function at the evaluation point specified in cell J2. In mathematical terms, this is computed as $\sum_{k=1}^{40} r_k$, where r_k denotes the kth cell of the “Real Product” row.

Excel Implementation 5.2.3.3

In the *Excel* program, the "Real Product" vector occupies C8:AP8. Therefore, this sum is implemented in *Excel* as:

$$= \text{SUM(C8:AP8)} \qquad (5.31)$$

6. IMAGINARY PART: Basis Functions at Evaluation Point. This is the imaginary part of each of the basis functions evaluated at the evaluation point given in cell J2. Let $z_0 = x_0 + iy_0$ represent the entry in J2. Recall the equation for the imaginary part of the CVBEM approximation function:

$$\hat{\psi}(x_0, y_0) = \sum_{j=1}^{n} \alpha_j \mathrm{Im}\left(\frac{1}{z_0 - z_j}\right) + \beta_j \mathrm{Re}\left(\frac{1}{z_0 - z_j}\right), \quad z_0 \in \Omega. \qquad (5.32)$$

Excel Implementation 5.2.3.4

Let z_j denote the location of the jth source point (i.e. the pole of the jth basis function), $1 \le j \le 20$. The first 20 cells in this row correspond to evaluating $\mathrm{Im}\left(\frac{1}{z_0 - z_j}\right)$. These cells are computed as follows:

$$= \text{IMAGINARY(IMDIV(1,IMSUB(\$J\$2,}z_j\text{)))} \qquad (5.33)$$

The next 20 cells in this row correspond to evaluating $\mathrm{Re}\left(\frac{1}{z_0 - z_j}\right)$. These cells are computed as follows:

$$= \text{IMAGINARY(IMPRODUCT(COMPLEX(0,1),...} \\ \text{IMDIV(1,IMSUB(\$J\$2,}z_j\text{))))} \qquad (5.34)$$

7. Imaginary Product. Again, let z_0 denote the evaluation point given in cell J2. Recall that the imaginary part of the CVBEM approximation function is of the form:

$$\hat{\psi}(x_0, y_0) = \sum_{j=1}^{n} \alpha_j \mathrm{Im}\left(\frac{1}{z_0 - z_j}\right) + \beta_j \mathrm{Re}\left(\frac{1}{z_0 - z_j}\right). \qquad (5.35)$$

This row is used to compute each of the products given by $\alpha_j \mathrm{Im}\left(\frac{1}{z_0 - z_j}\right)$ and $\beta_j \mathrm{Re}\left(\frac{1}{z_0 - z_j}\right)$ for each $1 \le j \le 20$.

***Excel* Implementation 5.2.3.5**

Define the following:

$$
\begin{aligned}
p_k &\equiv \text{the } k^{\text{th}} \text{ cell of the "Coefficients" row} \\
s_k &\equiv \text{the } k^{\text{th}} \text{ cell of the "IMAGINARY PART" row} \\
t_k &\equiv \text{the } k^{\text{th}} \text{ cell of the "Imaginary Product" row}
\end{aligned}
\tag{5.36}
$$

Given these variable names, the general equation for the "Imaginary Product" row is as follows:

$$t_k = \text{PRODUCT}(p_k, s_k) \tag{5.37}$$

8. Streamline Approximation. This is the CVBEM approximation of the stream function at the evaluation point specified in cell J2. In mathematical terms, this is computed as $\sum_{k=1}^{40} t_k$, where t_k denotes the kth cell of the "Imaginary Product" row.

***Excel* Implementation 5.2.3.6**

In the *Excel* program, the "Imaginary Product" vector occupies C11:AP11. Therefore, this sum is implemented in *Excel* as:

$$= \text{SUM(C11:AP11)} \tag{5.38}$$

9. dx and dy. These parameters are used to control the horizontal and vertical spacing between the domain evaluation points. In this case, we are using dx = dy = 0.05. The domain evaluation points form a grid where the number of rows and number of columns are calculated as follows:

$$
\begin{aligned}
\text{Number of rows } &= \frac{H}{\text{dy}} + 1 = \frac{1}{0.05} + 1 = 21 \\
\text{Number of columns } &= \frac{L}{\text{dx}} + 1 = \frac{1}{0.05} + 1 = 21,
\end{aligned}
\tag{5.39}
$$

where $H = 1$ is the vertical height of the problem domain, and $L = 1$ is the horizontal length of the problem domain. Thus, the domain evaluation points form a grid with 21 equally spaced rows of points and each row consisting of 21 equally spaced evaluation points. This means

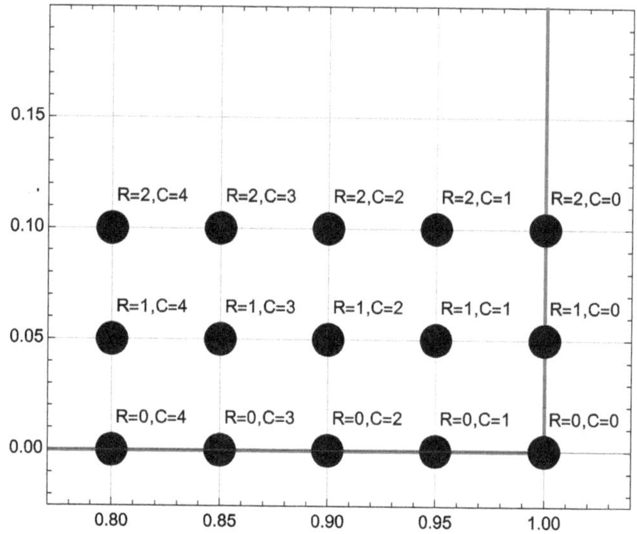

Figure 5.6: Depiction of the interior evaluation points near the bottom-right corner of the problem domain. In the figure, $R \equiv$ "row number" and $C \equiv$ "column number". Note: the rows are numbered such that the 0th row is on the bottom edge of the domain and the 20th row is on the top edge of the domain. The columns are numbered such that the 0th column is on the right edge of the domain and the 20th column is on the left edge of the domain. The columns are labeled in this seemingly counter-intuitive way so that when the data is transferred from the **Approximation Function** tab to the **Approximation** tab, the data is in the correct orientation to develop the desired contour plots.

there are a total of $21 \times 21 = 441$ domain evaluation points. A subset of these domain evaluation points near the bottom-right corner of the problem domain can be seen in Figure 5.15.

10. Column Counter. Since the domain evaluation points are spaced according to fixed values of dx and dy, the complex coordinate of each of these points can be determined by knowing the column number and row number of each point. The strategy is to use the "Point Number" column—which is displayed in Figure 5.7—to determine the column number and row number of each domain evaluation point.

    The location of each domain evaluation point is first determined for Row 0. Once the coordinates of all 21 of the points in Row 0 are computed, the

Point Number	Column Counter	Row Counter	Complex Coordinate	Potential Value		Complex Coordinate	Streamline Value
				0.00065851			-1.5471062
1	0	0	1	0.999669233		1	-1.5458437
2	1	0	0.95	0.902443981		0.95	-1.5458558
3	2	0	0.9	0.81011023		0.9	-1.5459388
4	3	0	0.85	0.72270586		0.85	-1.5460583
5	4	0	0.8	0.64023355		0.8	-1.5461866
6	5	0	0.75	0.56270394		0.75	-1.5463018
7	6	0	0.7	0.4901309		0.7	-1.5463873
8	7	0	0.65	0.42253075		0.65	-1.546432
9	8	0	0.6	0.35992144		0.6	-1.5464303
10	9	0	0.55	0.30232131		0.55	-1.5463825
→	→	→	→	→		→	→
431	10	20	0.5+i	-0.7500024		0.5+i	-0.5462181
432	11	20	0.45+i	-0.7975011		0.45+i	-0.6462176
433	12	20	0.4+i	-0.8399986		0.4+i	-0.7462221
434	13	20	0.35+i	-0.8774898		0.35+i	-0.8462304
435	14	20	0.3+i	-0.9099698		0.3+i	-0.9462382
436	15	20	0.25+i	-0.9374362		0.25+i	-1.0462375
437	16	20	0.2+i	-0.9598907		0.2+i	-1.1462179
438	17	20	0.15+i	-0.9773405		0.15+i	-1.2461681
439	18	20	0.1+i	-0.9897995		0.1+i	-1.3460783
440	19	20	)99999999999	-0.9972882	0.049999999999999		-1.4459433
441	20	20	i	-0.9998324		i	-1.5457657

Figure 5.7: Demonstration of the **Approximation Function** tab showing the first and last few rows of the dataset. The two red cells must be left empty in order to use the *Excel* Data Table program. The yellow cells are the input for the Data Table program. The blue cells reference the cells that compute the approximation of the potential and stream functions.

procedure is repeated for Row 1, Row 2, Row 3, etc. Therefore, the code in the "Column Counter" column needs to repeat the following pattern: $0, 1, \ldots, 20$, once for each of the 21 rows. A counter that repeatedly goes from 0 to 20 and then resets to 0 can be implemented using modular arithmetic as follows:

Excel Implementation 5.2.3.7

Define the following:

$$\text{point}_k \equiv \text{the } k^{\text{th}} \text{ cell of the "Point Number" column}$$
$$\text{col}_k \equiv \text{the } k^{\text{th}} \text{ cell of the "Column Counter" column}$$

Given these variable names, the column counter variable is computed as follows:

$$\text{col}_k = \text{IF(MOD(point}_k\text{,21)=0,21,MOD(point}_k\text{,21))}-1 \quad (5.40)$$

The if-statement is used to make sure that every 21st point in the "Point Number" column is labeled 20 in the "Column Counter" column. Without this if-statement, every 21st point in the "Point Number" column would be labeled −1 in the "Column Counter" column.

Note 5.2.3.1

An alternative to the equation given in *Excel* Implementation 5.2.3.7 for col_k that does not require an if-statement is as follows:

$$\text{col}_k = \text{MOD(point}_k - 1\text{,21)} \quad (5.41)$$

11. Row Counter. The row counter starts at 0 and remains at 0 for the first 21 points. Then, it is incremented by 1 and remains at 1 for the next 21 points. In this fashion, the row counter is incremented by 1 every 21 points. This is best implemented using the built-in FLOOR.MATH() function, which can be used to round a number down to the nearest integer. The implementation is as follows.

Excel Implementation 5.2.3.8

The FLOOR.MATH(number, significance) function has two arguments: the first is the number to be rounded, and the second is the multiple to which you want to round. For example, FLOOR.MATH(number, 2) rounds down to the nearest even number, while FLOOR.MATH(number, 3) rounds down to the nearest multiple of 3. For our purpose, we just want to round down to the nearest integer, so we use significance = 1.

Define the following:

$$\text{point}_k \equiv \text{the } k^{\text{th}} \text{ cell of the "Point Number" column}$$
$$\text{row}_k \equiv \text{the } k^{\text{th}} \text{ cell of the "Row Counter" column}$$

Given these variable names, the row counter variable is computed as follows:

$$\text{row}_k = \text{FLOOR.MATH}((\text{point}_k - 1)/21,1)$$

A good check is to compute this expression for "edge cases" such as the values $k = 21$ and $k = 22$. The first value corresponds to the last point of Row 0, and the second value corresponds to the first point of Row 1. So, we should get, $\text{row}_{21} = 0$ and $\text{row}_{22} = 1$. Observe:

$$\begin{aligned}\text{row}_{21} &= \text{FLOOR.MATH}((21-1)/21,1)\\ &= \text{FLOOR.MATH}(20/21,1)\\ &= \text{FLOOR.MATH}(0.95238,1)\\ &= 0\\ \text{row}_{22} &= \text{FLOOR.MATH}((22-1)/21,1)\\ &= \text{FLOOR.MATH}(21/21,1)\\ &= \text{FLOOR.MATH}(1,1)\\ &= 1\end{aligned} \tag{5.42}$$

12. Complex Coordinate. The complex coordinate of each domain evaluation point is computed based on the following data:

(a) Each point's column number and row number.

(b) The values of dx and dy.

(c) The length of the problem domain from the **Problem Data** tab.

***Excel* Implementation 5.2.3.9**

In this *Excel* application, the value of dx is stored in cell C14, and the value of dy is stored in cell C15. As before, let col_k denote the kth cell of the "Column Counter" column, and let row_k denote the kth cell of the "Row Counter" column. Let L denote the length of the problem domain from the **Problem Data** tab. Let z_k denote the complex coordinate of the kth domain evaluation point. Then,

$$z_k = \text{COMPLEX(L}-\text{row}_k\text{*\$C\$14,col}_k\text{*\$C\$15)} \qquad (5.43)$$

13. Potential Value and Streamline Value. These are the CVBEM approximations of the target potential function and stream function at each of the domain evaluation points. These approximations are computed using the "Data Table" feature of *Excel*. The instructions are as follows.

***Excel* Implementation 5.2.3.10**

For the potential value approximation:

(a) Identify the "Potential Approximation" cell, J3.

(b) In the blue cell of the "Potential Value" column, enter the formula =J3. The red cell must remain empty. Initially, all of the cells below the blue cell will be blank.

(c) Select the red cell, the blue cell, and all of the yellow cells (this will automatically select all of the empty cells below the blue cell that are adjacent to the yellow cells).

(d) In the "Data" tab of *Excel*, select "What-If Analysis".

(e) Select "Data Table...".

(f) Leave "Row input cell" blank.

(g) Enter $J$2 in "Column input cell".

(h) Click "OK". This will fill in the CVBEM approximations of the potential function at all of the domain evaluation points.

Excel Implementation 5.2.3.11

For the streamline value approximation:

(a) Identify the "Streamline Approximation" cell, X3.

(b) In the blue cell of the "Streamline Value" column, enter the formula =X3. The red cell must remain empty. Initially, all of the cells below the blue cell will be blank.

(c) Select the red cell, the blue cell, and all of the yellow cells (this will automatically select all of the empty cells below the blue cell that are adjacent to the yellow cells).

(d) In the "Data" tab of *Excel*, select "What-If Analysis".

(e) Select "Data Table...".

(f) Leave "Row input cell" blank.

(g) Enter $J$2 in "Column input cell".

(h) Click "OK". This will fill in the CVBEM approximations of the stream function at all of the domain evaluation points.

5.2.4 The Approximation tab

As discussed in Section 5.1.4, the **Approximation** tab consists of two matrices:

1. The first matrix consists of values obtained by evaluating the real part of the CVBEM approximation function at various points in the problem domain. These values are used to produce the contour plot of potential values in the **Plots** tab.

2. The second matrix consists of values obtained by evaluating the imaginary part of the CVBEM approximation function at the same points in the problem domain. These values are used to produce the contour plot of streamline values in the **Plots** tab.

The **Approximation Tab** is updated automatically and does not require user interaction.

5.2.5 The Plots tab

The **Plots** tab is used to visualize the computational results from the CVBEM model. Specifically, in Figure 5.2 at the beginning of this section, we have depicted the locations of the nodes and collocation points used in the CVBEM model. In Figure 5.8, we have depicted contours from the CVBEM approximation of the potential function. And in Figure 5.9, we have depicted contours from the CVBEM approximation of the stream function. These figures have been developed according to the instructions listed in Section 5.1.5.

Figure 5.8: Contours produced in *Excel* from the real part (potential function) of the CVBEM approximation function. These contours are from a CVBEM model of potential flow in a 90-degree bend.

Figure 5.9: Contours produced in *Excel* from the imaginary part (stream function) of the CVBEM approximation function. These contours are from a CVBEM model of potential flow in a 90-degree bend.

5.3 Potential flow over a cylindrical obstacle

In this section, we consider another benchmark problem; namely, potential flow over a cylindrical obstacle. An interesting feature of this problem is that the target solution is symmetric about both the horizontal and vertical axes, as illustrated in Figure 5.10. That is, the solution in quadrants II, III, and IV can be obtained from the solution in quadrant I by either 1 or 2 reflections. Hence, the size of the problem domain can be reduced to focus on one half—or even one quarter—of the area of interest. In this case, we will focus on modeling the upper half of the area of interest, $\text{Im}(z) \geq 0$.

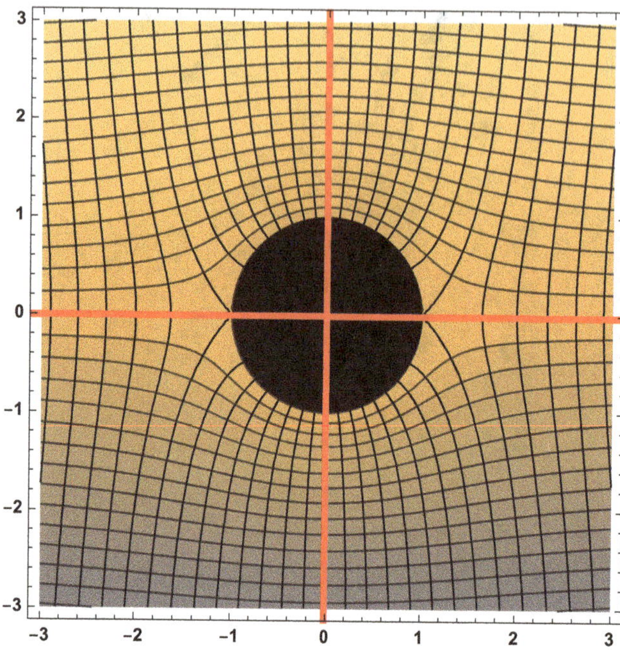

Figure 5.10: Analytic solution to the problem of potential flow around a circular cylinder. In this case, the target solution has symmetry about both the horizontal and vertical axes, as indicated by the red lines. This symmetry means we can reduce the problem domain to either one half or one quarter of the area of interest.

Note 5.3.0.1

The reason for modeling the upper half-plane as opposed to just one of the quadrants has to do with our desire to remain consistent with other work we have done on this problem, such as in [29]. In general, one of the advantages of modeling this problem in two quadrants is that it enables a particular kind of error-checking. Specifically, we know that the solution in quadrant II should be a reflection of the solution in quadrant I. Therefore, given that this should be the case, by modeling both quadrants I and II, we can see how well the CVBEM approximation function actually achieves this symmetry.

This idea of checking the symmetry of the CVBEM solution is not evaluated further in this book. However, hopefully the inspiration for choosing to model the upper half-plane as opposed to just one of the quadrants is more clear.

5.3.1 The Problem Data tab

In this example, the problem domain is defined as follows:

$$\Omega = \Omega_1 \setminus \Omega_2 \quad \text{where } \Omega_1 = [-3,3] \times [0,3] \text{ and } \Omega_2 = x^2 + y^2 \leq 1. \tag{5.44}$$

Given this definition, we specify the Length parameter as 6 and the Height parameter as 2, as indicated in Table 5.2. In Section 5.2, the example problem considered a problem geometry in which the bottom left corner of the problem domain was located at the origin. Now, we consider a problem geometry in which this is no longer the case. Therefore, the **Problem Data** tab has two new fields to address this situation; namely "Bottom left corner of Ω (x-coordinate)" and "Bottom left corner of Ω (y-coordinate)". These two fields, and their appropriate values, are shown in Table 5.2.

In this problem, we are going to create a CVBEM model with $n = 20$ terms in the approximation function. Therefore, this requires the specification of $n = 20$ computational nodes and $2n = 40$ collocation points.

In the previous problem, the locations of the computational nodes and collocation points were prescribed according to a regular geometric pattern. In particular, the collocation points were evenly spaced along the problem boundary and the computational nodes were evenly spaced along a circle enveloping the problem domain. This approach is usually easy to implement; however, the resulting CVBEM model is not usually optimal. To address this problem, node positioning algorithms (NPAs) have been developed that seek to determine better locations for the collocation points and computational nodes. These NPAs have been described in the recent papers [30, 29].

Note 5.3.1.1

Using an NPA to determine the locations of collocation points and computational nodes often results in a more accurate model than the one obtained using the geometric approach described in Section 5.2. However, even though the resulting model is often more accurate, there is still not any guarantee of optimality, and it is often the case that the models resulting from the use of an NPA can be demonstrated to be sub-optimal.

Nevertheless, the improved accuracy afforded by the use of an NPA is often sufficient for most analyses.

In this problem, we demonstrate a CVBEM model in which the placement of collocation points and computational nodes has been determined with the use of an NPA. As such, the collocation points and computational nodes do not appear in a regular pattern. Consequently, the locations of the computational nodes cannot be defined by a single parameter, such as dtheta, as in the previous problem demonstrated in Section 5.2.

Note 5.3.1.2

The mechanics of the NPA are not strictly related to the study and application of complex variables. For this reason, a detailed description of the NPA is outside the scope of this book. However, for more details on the algorithms used to determine the locations of the collocation points and nodes used in the CVBEM model, the interested reader is referred to [30, 29], in which two generations of the NPA are discussed.

***Excel* Implementation 5.3.1.1**

Length	6
Height	3
Bottom left corner of Ω (x-coordinate)	−3
Bottom left corner of Ω (y-coordinate)	0

Table 5.2: The user will input problem data into a table of this form in the **Problem Data** tab. The two fields related to the bottom-left corner are used to indicate the coordinates of the bottom-left corner of the problem domain. The four parameters shown in this table uniquely define the problem domain.

This setup is depicted in Figure 5.11 and can be found in the **Plots** tab of the accompanying *Excel* workbook.

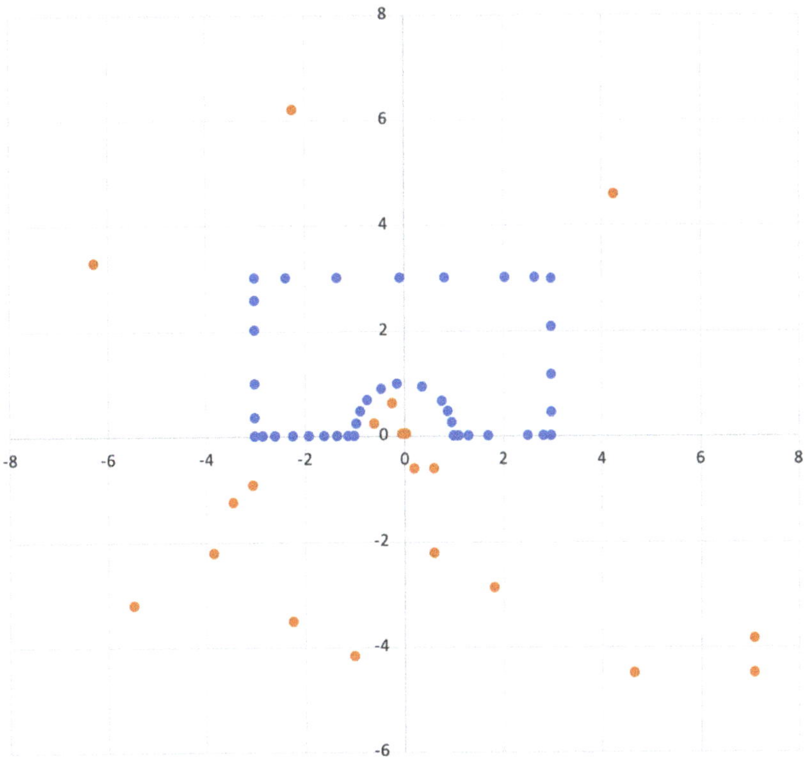

Figure 5.11: Locations of nodes and collocation points used in the CVBEM model. Nodes are depicted as orange points, and collocation points are depicted as blue points. The physical phenomena being modeled is potential flow over a cylinder.

The "Boundary Condition" column of the **Problem Data** tab is found by evaluating the target potential function at each of the collocation points. For this BVP, the boundary conditions are given as:

$$\phi(x, y) = \mathrm{Re}\left(z + \frac{1}{z}\right) = x + \frac{x}{x^2 + y^2}, \quad (x, y) \in \partial\Omega. \tag{5.45}$$

Therefore, the boundary conditions are computed as $x_k + \frac{x_k}{x_k^2 + y_k^2}$, where $(x_k, y_k) \in \partial\Omega$ denotes the location of the kth boundary condition.

Excel Implementation 5.3.1.2

Figures 5.13 and 5.14 show the complex coordinates for the computational nodes used in the CVBEM approximation function for this problem. Since the basis functions used in this model are of the form $\frac{1}{z-z_j}$, the computational nodes are identified as the poles of the basis functions.

We will give the general equation for each row shown in these figures. Define the following:

$$\begin{aligned} x_k &\equiv \text{the } k^{\text{th}} \text{ cell of the } x\text{-coordinate row} \\ y_k &\equiv \text{the } k^{\text{th}} \text{ cell of the } y\text{-coordinate row} \\ z_k &\equiv \text{the } k^{\text{th}} \text{ cell of the complex coordinate row} \end{aligned} \tag{5.46}$$

Given these variable names, the general equations for each of the rows shown in Figures 5.13 and 5.14 are as follows:

1. x-coordinate of poles: MANUALLY ENTERED (output from use of NPA).
2. y-coordinate of poles: MANUALLY ENTERED (output from use of NPA).
3. Complex coordinate of poles: z_i = COMPLEX(x_i, y_i).

The manually entered data in this implementation were the result of the use of an NPA. However, in practice, a user may select any locations for the computational nodes as long as the locations are exterior to the problem domain.

Excel Implementation 5.3.1.3

The table shown in Figure 5.12 contains the complex coordinates of the collocation points as well as the value of the boundary condition at each of the collocation points. The first collocation point is located at the origin, $(-3,0)$. Then, the collocation points are placed counterclockwise along the problem domain. In this example, the spacing of the collocation points is not uniform. Define the following:

$$\begin{aligned} x_k &\equiv \text{the } k^{\text{th}} \text{ cell of the } x\text{-coordinate column} \\ y_k &\equiv \text{the } k^{\text{th}} \text{ cell of the } y\text{-coordinate column} \\ z_k &\equiv \text{the } k^{\text{th}} \text{ cell of the complex coordinate column} \\ \phi_k &\equiv \text{the } k^{\text{th}} \text{ cell of the boundary condition column} \end{aligned} \tag{5.47}$$

Given these variable names, the general equations for the complex coordinate column and the boundary condition column shown in Figure 5.12 is as follows:

1. Complex coordinate column: $z_k = \text{COMPLEX}(x_k, y_k)$
2. Boundary condition column:

$$\phi_k = \text{IMREAL(IMSUM(}z_k\text{,IMDIV(COMPLEX(0,1),}z_k\text{)))}$$

The equation for the boundary condition column is defined by the boundary conditions of the problem. In this case, the boundary conditions are $\phi(x,y) = \text{Re}\left(z + \frac{1}{z}\right)$, $(x,y) \in \partial\Omega$.

x-coordinate of collocation points	y-coordinate of collocation points	Complex coordinate of collocation points	Boundary Condition
-3	0	-3	-3.333333333
3	2.982	2.9999999999998	3.167669667
-2.994	3	-2.994000000000(	-3.160666333
2.994	0	2.9939999999999	3.328001336
1	0	0.9999999999998	2
-1	0	-1	-2
1.698	0	1.6979999999999	2.286928151
0.755749574	0.654860734	0.7557495743539	1.511499149
-0.736741138	0.6761749	-0.736741137876(	-1.473482276
-3	0.99	-3+0.9900000000(	-3.30059819
-0.142314838	0.989821442	-0.142314838272!	-0.284629677
3	1.164	2.9999999999998	3.289718023
-1.902	0	-1.901999999999!	-2.427762355
2.046	3	2.0459999999999	2.201163203
-1.344	0	-1.344000000000(	-2.088047619
2.508	0	2.5079999999999	2.906724083
3	0.444	2.9999999999998	3.3261885
3	2.082	2.9999999999998	3.224976535
-1.356	3	-1.356000000000(	-1.481106839
-3	2.016	-3+2.0160000000(	-3.229634202
-2.364	3	-2.364000000000(	-2.526045491
-2.604	0	-2.604000000000(	-2.988024578
0.362807705	0.931864029	0.3628077053510	0.725615411
-0.078	3	-0.078000000000(	-0.086660812
1.302	0	1.3019999999999	2.070049155
-3	0.342	-3+0.3420000000(	-3.32905691
0.967146855	0.254218334	0.9671468547018	1.934293709
-0.974011917	0.226496767	-0.974011916942(	-1.948023834
-3	2.574	-3+2.5740000000(	-3.191994151
-0.466667323	0.884432931	-0.466667323225!	-0.933334646
0.884432931	0.466667323	0.8844329309980	1.768865862
0.822	3	0.8220000000001	0.906955234
-2.226	0	-2.225999999999!	-2.675236298
-1.614	0	-1.614	-2.233578686
2.658	3	2.6580000000000	2.823453219
-1.128	0	-1.128000000000(	-2.014524823
-2.838	0	-2.838	-3.190360817
1.104	0	1.1039999999999	2.009797101
-0.884432931	0.466667323	-0.884432930998(	-1.768865862
2.832	0	2.8319999999999	3.185107345

Figure 5.12: Table in the *Excel* program specifying the x- and y-coordinates of the collocation points. In the third column, the real coordinates of the collocation points are converted to the corresponding point in the complex plane. In the final column, we specify the value of the boundary condition at the particular collocation point. For this problem, the boundary conditions are $\phi_k(x_k, y_k) = \text{Re}\left(z_k + \frac{1}{z_k}\right)$, $(x_k, y_k) \in \partial\Omega$.

x-coordinate of poles:	7.095	-5.473	4.662	7.095	-0.047	0.047	-1.014	0.608	1.824	4.257
y-coordinate of poles:	-3.851	-3.203	-4.500	-4.500	0.047	0.047	-4.176	-2.230	-2.878	4.581
Complex coordinate of poles	7.0-3.8i	-5.4-3.2i	4.6-4.5i	7.0-4.5i	-0.0+0.0i	0.0+0.0i	-1.0-4.1i	0.6-2.2i	1.8-2.8i	4.2+4.5i

Figure 5.13: Table in *Excel* specifying the x- and y-coordinates of the computational nodes. In the third row, the real coordinates (shown in the first and second rows) of the computational nodes are converted to the corresponding point in the complex plane.

x-coordinate of poles:	-3.851	0.203	-2.230	-0.236	-2.230	0.608	-3.041	-3.446	-6.284	-0.613
y-coordinate of poles:	-2.230	-0.608	-3.527	0.613	6.203	-0.608	-0.932	-1.257	3.284	0.236
Complex coordinate of poles	-3.8-2.2i	0.2-0.6i	-2.2-3.5i	-2.2+0.6i	-2.2+6.2i	0.6-0.6i	-3.0-0.9i	-3.4-1.2i	-6.2+3.2i	-0.6+0.2i

Figure 5.14: Continuation of the data presented in Figure 5.13.

5.3.2 The Coefficient Matrix tab

The **Coefficient Matrix** tab consists of the following parts:

1. $\boxed{A \in \mathbb{R}^{40\times 40}.}$ This matrix contains the real and imaginary parts of each of the 20 basis functions evaluated at each of the 40 collocation points. It is best to think of this matrix in block form, as follows:

$$A = \begin{bmatrix} \underbrace{A_1}_{\substack{\text{The real} \\ \text{part of the} \\ \text{basis functions}}} & \underbrace{A_2}_{\substack{\text{The imaginary} \\ \text{part of the} \\ \text{basis functions}}} \end{bmatrix}, \quad A_1, A_2 \in \mathbb{R}^{40\times 20}. \tag{5.48}$$

Let (x_i, y_i) for $i = 1, \ldots, 40$ denote the locations of the collocation points. Then, $A_1(i,j) = \lambda_j(x_i, y_i)$ and $A_2(i,j) = -\mu_j(x_i, y_i)$ (these expressions are based on the equation for $\hat{\phi}(x,y)$ in Equation (5.3)).

$$\begin{aligned} A_1 &= \begin{bmatrix} \lambda_1(x_1,y_1) & \lambda_2(x_1,y_1) & \cdots & \lambda_{20}(x_1,y_1) \\ \lambda_1(x_2,y_2) & \lambda_2(x_2,y_2) & \cdots & \lambda_{20}(x_2,y_2) \\ \vdots & \vdots & \ddots & \vdots \\ \lambda_1(x_{40},y_{40}) & \lambda_2(x_{40},y_{40}) & \cdots & \lambda_{20}(x_{40},y_{40}) \end{bmatrix} \\ A_2 &= \begin{bmatrix} -\mu_1(x_1,y_1) & -\mu_2(x_1,y_1) & \cdots & -\mu_{20}(x_1,y_1) \\ -\mu_1(x_2,y_2) & -\mu_2(x_2,y_2) & \cdots & -\mu_{20}(x_2,y_2) \\ \vdots & \vdots & \ddots & \vdots \\ -\mu_1(x_{40},y_{40}) & -\mu_2(x_{40},y_{40}) & \cdots & -\mu_{20}(x_{40},y_{40}) \end{bmatrix} \end{aligned} \tag{5.49}$$

There are at least two ways to calculate $-\mu_j(x_i, y_i)$. One way is as $-\text{Im}\left(\frac{1}{z_i - z_j}\right)$, where

$$\begin{aligned} z_i &= x_i + iy_i \equiv \text{ the } i^{\text{th}} \text{ collocation point} \\ z_j &= x_j + iy_j \equiv \text{ the } j^{th} \text{ computational node} \end{aligned} \tag{5.50}$$

Note: the difference between the index i and the imaginary number should be clear from the context. The alternative way is as $\text{Re}\left(\frac{i}{z_i - z_j}\right)$. We use the latter option in the *Excel* implementation demonstrated in this work.

***Excel* Implementation 5.3.2.1**

A general cell of the matrix $A = \begin{bmatrix} A_1 & A_2 \end{bmatrix}$ is computed as follows:

$$A_1(i,j) = \text{IMREAL(IMDIV(1,IMSUB(}z_i\text{,}z_j\text{)))}$$
$$A_2(i,j) = \text{IMREAL(IMPRODUCT(COMPLEX(0,1),...} \quad (5.51)$$
$$\text{IMDIV(1,IMSUB(}z_i\text{,}z_j\text{))))}$$

2. $x \in \mathbb{R}^{20}$. This vector consists of the coefficients of the CVBEM approximation function. The user has to initialize this vector. A choice of all 1's is usually sufficient. Then, the built-in "Solver" application is used to determine the values of these coefficients by optimization. The "Solver" application can be configured as shown in Figure 5.1.

3. $Ax = \hat{b} \in \mathbb{R}^{20}$. This vector contains the CVBEM approximation of the target boundary conditions.

***Excel* Implementation 5.3.2.2**

The CVBEM approximation of the target boundary conditions is computed using the MMULT() function. In this case, the A matrix is stored in the cells A2:AN41 and the x vector is stored in the cells AO2:AO41. Therefore, $\hat{b}$ is calculated as:

$$\hat{b} = \{\text{ MMULT(A2:AN41,AO2:AO41) }\} \quad (5.52)$$

Note: the *Excel* expression for $\hat{b}$ is surrounded by curly braces. This syntax denotes that a matrix operation is being used. In order to perform a matrix operation like this, it is necessary to enter the command slightly differently. The following instructions can be used:

(a) Select the region where the results are to be stored. In this case, the results are to be stored in the column labeled $Ax = \hat{b}$. Therefore, we select the cells AP2:AP41.

(b) Enter the following: = MMULT(A2:AN41,AO2:AO41).

(c) Here is the important step. After entering the command, it is not enough to just press "Enter" on the keyboard. Instead, it is necessary to press "Control + Shift + Enter".

4. $b \in \mathbb{R}^{20}$. This vector contains the target boundary conditions. The data in this vector is automatically copied from the "Boundary Condition" column in the **Problem Data** tab.

5. Absolute error. This vector is the absolute difference between $\hat{b}$ and b, which is computed as $\varepsilon = \left|\hat{b} - b\right|$.

***Excel* Implementation 5.3.2.3**

Define the following:

$$\begin{aligned} \hat{b}_k &\equiv \text{the } k^{\text{th}} \text{ cell of the } Ax = \hat{b} \text{ column} \\ b_k &\equiv \text{the } k^{\text{th}} \text{ cell of the } b \text{ column} \\ \varepsilon_k &\equiv \text{the } k^{\text{th}} \text{ cell of the Absolute Error column} \end{aligned} \tag{5.53}$$

Given these variable names, the general equation for the cells in the Absolute Error column is as follows:

$$\varepsilon_k = \text{ABS}(\hat{b}_k - b_k) \tag{5.54}$$

6. Error Squared. This vector contains the square of each element in ε.

***Excel* Implementation 5.3.2.4**

Define the following:

$$\begin{aligned} \varepsilon_k &\equiv \text{the } k^{\text{th}} \text{ cell of the Absolute Error column} \\ \varepsilon_k^2 &\equiv \text{the } k^{\text{th}} \text{ cell of the Error Squared column} \end{aligned} \tag{5.55}$$

Given these variable names, the general equation for the cells in the Error Squared column is as follows:

$$\varepsilon_k^2 = \text{POWER}(\varepsilon_k,2) \tag{5.56}$$

7. Sum of Squared Errors. This cell is equivalent to $\varepsilon^{\top}\varepsilon$ and to $\sum_k \varepsilon_k^2$. This cell is used as the objective cell for the "Solver" application.

Excel Implementation 5.3.2.5

Let SSE denote the sum of squared errors. For this example problem, the ε^2 vector is stored in the cells AT2:AT41. Therefore, the Sum of Squared Errors cell is calculated as follows:

$$\text{SSE} = \text{SUM(AT2:AT41)} \tag{5.57}$$

5.3.3 The Approximation Function tab

The **Approximation Function** tab is used to evaluate the CVBEM approximation function at various points in the interior of the problem domain. These evaluated values are used to create the contour plots in the **Plots** tab. The features of this tab are as follows:

1. Evaluation point: z =. This is the point at which the CVBEM approximation function is to be evaluated. It is used as the "Column input cell" in *Excel*'s built-in "Data Table" program.

2. Coefficients. These are the coefficients that were obtained by optimization in the **Coefficient Matrix** tab.

3. REAL PART: Basis Functions at Evaluation Point. This is the real part of each of the basis functions evaluated at the evaluation point given in cell J2. Let $z_0 = x_0 + iy_0$ represent the entry in J2. Recall the equation for the real part of the CVBEM approximation function:

$$\hat{\phi}(x_0, y_0) = \sum_{j=1}^{n} \alpha_j \text{Re}\left(\frac{1}{z_0 - z_j}\right) - \beta_j \text{Im}\left(\frac{1}{z_0 - z_j}\right), \quad z_0 \in \Omega. \tag{5.58}$$

Excel Implementation 5.3.3.1

Let z_j denote the location of the jth source point (i.e. the pole of the jth basis function), $1 \le j \le 20$. The first 20 cells in this row correspond to evaluating $\text{Re}\left(\frac{1}{z_0 - z_j}\right)$. These cells are computed as follows:

$$= \text{IMREAL(IMDIV(1,IMSUB(\$J\$2,}z_j\text{)))} \qquad (5.59)$$

The next 20 cells in this row correspond to evaluating $-\text{Im}\left(\frac{1}{z_0 - z_j}\right)$. These cells are computed as follows:

$$= \text{IMREAL(IMPRODUCT(COMPLEX(0,1),...} \\ \text{IMDIV(1,IMSUB(\$J\$2,}z_j\text{))))} \qquad (5.60)$$

4. Real Product. Again, let z_0 denote the evaluation point given in cell J2. Recall that the real part of the CVBEM approximation function is of the form:

$$\hat{\phi}(x_0, y_0) = \sum_{j=1}^{n} \alpha_j \text{Re}\left(\frac{1}{z_0 - z_j}\right) - \beta_j \text{Im}\left(\frac{1}{z_0 - z_j}\right). \qquad (5.61)$$

This row is used to compute each of the products given by $\alpha_j \text{Re}\left(\frac{1}{z_0 - z_j}\right)$ and $-\beta_j \text{Im}\left(\frac{1}{z_0 - z_j}\right)$ for each $1 \le j \le 20$.

***Excel* Implementation 5.3.3.2**

Define the following:

$$\begin{aligned} p_k &\equiv \text{the } k^{\text{th}} \text{ cell of the "Coefficients" row} \\ q_k &\equiv \text{the } k^{\text{th}} \text{ cell of the "REAL PART" row} \\ r_k &\equiv \text{the } k^{\text{th}} \text{ cell of the "Real Product" row} \end{aligned} \qquad (5.62)$$

Given these variable names, the general equation for the "Real Product" row is as follows:

$$r_k = \text{PRODUCT}(p_k, q_k) \qquad (5.63)$$

5. Potential Approximation. This is the CVBEM approximation of the potential function at the evaluation point specified in cell J2. In mathematical terms, this is computed as $\sum_{k=1}^{40} r_k$, where r_k denotes the kth cell of the "Real Product" row.

***Excel* Implementation 5.3.3.3**

In the *Excel* program, the "Real Product" vector occupies C8:AP8. Therefore, this sum is implemented in *Excel* as:

$$= \text{SUM(C8:AP8)} \quad (5.64)$$

6. IMAGINARY PART: Basis Functions at Evaluation Point. This is the imaginary part of each of the basis functions evaluated at the evaluation point given in cell J2. Let $z_0 = x_0 + iy_0$ represent the entry in J2. Recall the equation for the imaginary part of the CVBEM approximation function:

$$\hat{\psi}(x_0, y_0) = \sum_{j=1}^{n} \alpha_j \text{Im}\left(\frac{1}{z_0 - z_j}\right) + \beta_j \text{Re}\left(\frac{1}{z_0 - z_j}\right), \quad z_0 \in \Omega. \quad (5.65)$$

***Excel* Implementation 5.3.3.4**

Let z_j denote the location of the jth source point (i.e. the pole of the jth basis function), $1 \le j \le 20$. The first 20 cells in this row correspond to evaluating $\text{Im}\left(\frac{1}{z_0 - z_j}\right)$. These cells are computed as follows:

$$= \text{IMAGINARY(IMDIV(1,IMSUB(\$J\$2},z_j\text{)))} \quad (5.66)$$

The next 20 cells in this row correspond to evaluating $\text{Re}\left(\frac{1}{z_0 - z_j}\right)$. These cells are computed as follows:

$$\begin{aligned} &= \text{IMAGINARY(IMPRODUCT(COMPLEX(0,1),...} \\ &\quad \text{IMDIV(1,IMSUB(\$J\$2},z_j\text{))))} \end{aligned} \quad (5.67)$$

7. Imaginary Product. Again, let z_0 denote the evaluation point given in cell J2. Recall that the imaginary part of the CVBEM approximation function is of the form:

$$\hat{\psi}(x_0, y_0) = \sum_{j=1}^{n} \alpha_j \text{Im}\left(\frac{1}{z_0 - z_j}\right) + \beta_j \text{Re}\left(\frac{1}{z_0 - z_j}\right). \quad (5.68)$$

This row is used to compute each of the products given by $\alpha_j \text{Im}\left(\frac{1}{z_0 - z_j}\right)$ and $\beta_j \text{Re}\left(\frac{1}{z_0 - z_j}\right)$ for each $1 \le j \le 20$.

***Excel* Implementation 5.3.3.5**

Define the following:

$$\begin{aligned} p_k &\equiv \text{the } k^{\text{th}} \text{ cell of the "Coefficients" row} \\ s_k &\equiv \text{the } k^{\text{th}} \text{ cell of the "IMAGINARY PART" row} \\ t_k &\equiv \text{the } k^{\text{th}} \text{ cell of the "Imaginary Product" row} \end{aligned} \tag{5.69}$$

Given these variable names, the general equation for the "Imaginary Product" row is as follows:

$$t_k = \text{PRODUCT}(p_k, s_k) \tag{5.70}$$

8. Streamline Approximation. This is the CVBEM approximation of the stream function at the evaluation point specified in cell J2. In mathematical terms, this is computed as $\sum_{k=1}^{40} t_k$, where t_k denotes the kth cell of the "Imaginary Product" row.

***Excel* Implementation 5.3.3.6**

In the *Excel* program, the "Imaginary Product" vector occupies C11:AP11. Therefore, this sum is implemented in *Excel* as:

$$= \text{SUM(C11:AP11)} \tag{5.71}$$

9. dx and dy. These parameters are used to control the horizontal and vertical spacing between the domain evaluation points. In this case, we are using dx = 0.12 and dy = 0.06. The domain evaluation points form a grid where the number of rows and number of columns are calculated as follows:

$$\begin{aligned} \text{Number of rows} &= \frac{H}{\text{dy}} + 1 = \frac{3}{0.06} + 1 = 51 \\ \text{Number of columns} &= \frac{L}{\text{dx}} + 1 = \frac{6}{0.12} + 1 = 51, \end{aligned} \tag{5.72}$$

where $H = 3$ is the vertical height of the problem domain, and $L = 6$ is the horizontal length of the problem domain. Thus, the domain evaluation points form a grid with 51 equally spaced rows of points and each row consisting of 51 equally spaced evaluation points. This means

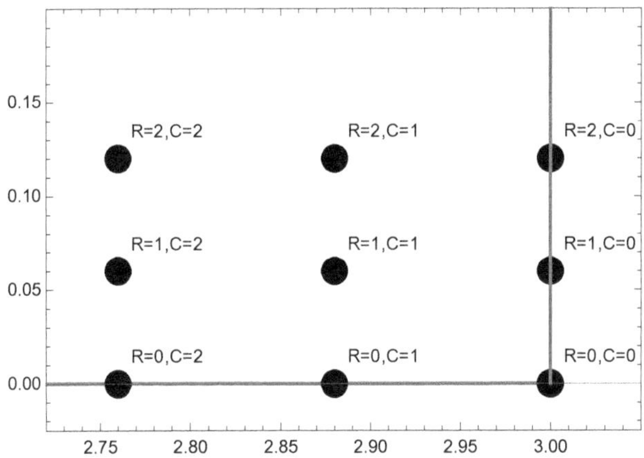

Figure 5.15: Depiction of the interior evaluation points near the bottom-right corner of the problem domain. In the figure, $R \equiv$ "row number" and $C \equiv$ "column number". Note: the rows are numbered such that the 0th row is on the bottom edge of the domain and the 20th row is on the top edge of the domain. The columns are numbered such that the 0th column is on the right edge of the domain and the 20th column is on the left edge of the domain. The columns are labeled in this seemingly counter-intuitive way so that when the data is transferred from the **Approximation Function** tab to the **Approximation** tab, the data is in the correct orientation to develop the desired contour plots.

there are a total of $51 \times 51 = 2601$ domain evaluation points. A subset of these domain evaluation points near the bottom-right corner of the problem domain can be seen in Figure 5.15.

Note 5.3.3.1

Using a rectangular grid to determine the locations of the domain evaluation points is convenient. However, this will result in some domain evaluation points being located in the area of the cylinder, which is not part of the problem domain. In this case, we will still evaluate the CVBEM approximation function at these points, and we will still plot them as part of the resulting contour plots. However, they will be covered by the cylinder graphic, so they will not be seen.

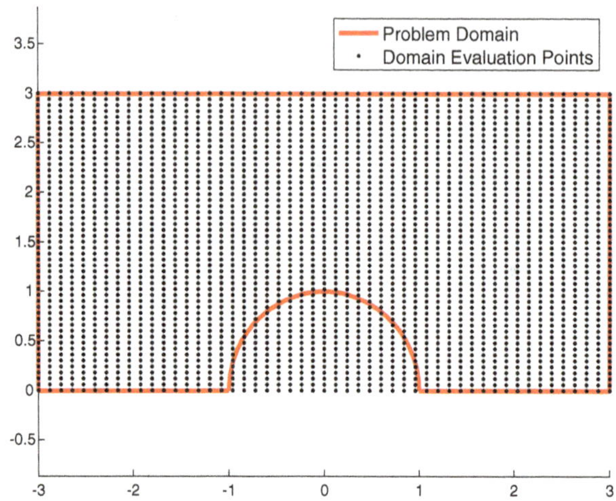

Figure 5.16: Locations of the domain evaluation points for this problem. Because we are using a rectangular grid to define the locations of the domain evaluation points, some of the points are located in the exterior of the problem domain. In particular, some domain evaluation points are located in the area of the cylinder obstruction. This is not a problem, and the cylinder graphic will cover the contour plot that is produced in this area.

10. Column Counter. Since the domain evaluation points are spaced according to fixed values of dx and dy, the complex coordinate of each of these points can be determined by knowing the column number and row number of each point. The strategy is to use the "Point Number" column—which is displayed in Figure 5.17—to determine the column number and row number of each domain evaluation point.

    The location of each domain evaluation point is first determined for Row 0. Once the coordinates of all 51 of the points in Row 0 are computed, the procedure is repeated for Row 1, Row 2, Row 3, etc. Therefore, the code in the "Column Counter" column needs to repeat the following pattern: $0, 1, \ldots, 50$, once for each of the 51 rows. A counter that repeatedly goes from 0 to 50 and then resets to 0 can be implemented using modular arithmetic as follows:

Point Number	Column Counter	Row Counter	Complex Coordinate	Potential Value		Complex Coordinate	Streamline Value
				0.43703878			-25.399335
1	0	0	3	3.33333426		3	-40.699292
2	1	0	2.88	3.22721854		2.88	-40.699271
3	2	0	2.76	3.12232584		2.76	-40.699267
4	3	0	2.64	3.01880061		2.64	-40.69928
5	4	0	2.52	2.91682743		2.52	-40.699298
6	5	0	2.4	2.81664281		2.4	-40.699306
7	6	0	2.28	2.71854314		2.28	-40.699295
8	7	0	2.16	2.62289141		2.16	-40.699266
9	8	0	2.04	2.53012717		2.04	-40.699232
10	9	0	1.92	2.44078634		1.92	-40.699207
↓	↓	↓	↓	↓		↓	↓
2591	40	50	-1.8+3i	-1.9470415		-1.8+3i	-37.944364
2592	41	50	-1.92+3i	-2.0713243		-1.92+3i	-37.935734
2593	42	50	-2.04+3i	-2.1949786		-2.04+3i	-37.92719
2594	43	50	-2.16+3i	-2.3180475		-2.16+3i	-37.918777
2595	44	50	-2.28+3i	-2.440575		-2.28+3i	-37.910535
2596	45	50	-2.4+3i	-2.5626047		-2.4+3i	-37.902495
2597	46	50	-2.52+3i	-2.6841786		-2.52+3i	-37.894681
2598	47	50	-2.64+3i	-2.8053366		-2.64+3i	-37.887113
2599	48	50	-2.76+3i	-2.9261155		-2.76+3i	-37.879805
2600	49	50	-2.88+3i	-3.0465486		-2.88+3i	-37.872763
2601	50	50	-3+3i	-3.166665		-3+3i	-37.865991

Figure 5.17: Demonstration of the **Approximation Function** tab showing the first and last few rows of the dataset. The two red cells must be left empty in order to use the *Excel* Data Table program. The yellow cells are the input for the Data Table program. The blue cells reference the cells that compute the approximation of the potential and stream functions.

***Excel* Implementation 5.3.3.7**

Define the following:

$$\text{point}_k \equiv \text{the } k^{\text{th}} \text{ cell of the "Point Number" column}$$
$$\text{col}_k \equiv \text{the } k^{\text{th}} \text{ cell of the "Column Counter" column}$$

Given these variable names, the column counter variable is computed as follows:

$$\text{col}_k = \text{IF(MOD(point}_k\text{,51)=0,51,MOD(point}_k\text{,51))}-1 \quad (5.73)$$

The if-statement is used to make sure that every 51st point in the "Point Number" column is labeled 50 in the "Column Counter" column. Without this if-statement, every 51st point in the "Point Number" column would be labeled −1 in the "Column Counter" column.

Note 5.3.3.2

An alternative to the equation given in Excel Implementation 5.3.3.7 for col_k that does not require an if-statement is as follows:

$$\text{col}_k = \text{MOD(point}_k - 1\text{,21)} \quad (5.74)$$

11. Row Counter. The row counter starts at 0 and remains at 0 for the first 51 points. Then, it is incremented by 1 and remains at 1 for the next 51 points. In this fashion, the row counter is incremented by 1 every 51 points. This is best implemented using the built-in FLOOR.MATH() function, which can be used to round a number down to the nearest integer. The implementation is as follows.

***Excel* Implementation 5.3.3.8**

The FLOOR.MATH(number, significance) function has two arguments: the first is the number to be rounded, and the second is the multiple to which you want to round. For example, FLOOR.MATH(number, 2) rounds down to the nearest even number, while FLOOR.MATH(number, 3) rounds down to the

nearest multiple of 3. For our purpose, we just want to round down to the nearest integer, so we use significance = 1.
Define the following:

$$\text{point}_k \equiv \text{the } k^{\text{th}} \text{ cell of the “Point Number” column}$$
$$\text{row}_k \equiv \text{the } k^{\text{th}} \text{ cell of the “Row Counter” column}$$

Given these variable names, the row counter variable is computed as follows:

$$\text{row}_k = \text{FLOOR.MATH}((\text{point}_k - 1)/51,1) \tag{5.75}$$

A good check is to compute this expression for “edge cases” such as the values $k = 51$ and $k = 52$. The first value corresponds to the last point of Row 0, and the second value corresponds to the first point of Row 1. So, we should get, $\text{row}_{51} = 0$ and $\text{row}_{52} = 1$. Observe:

$$\begin{aligned}\text{row}_{51} &= \text{FLOOR.MATH}((51 - 1)/51,1)\\ &= \text{FLOOR.MATH}(50/51,1)\\ &= \text{FLOOR.MATH}(0.98039,1)\\ &= 0\\ \text{row}_{22} &= \text{FLOOR.MATH}((52 - 1)/51,1)\\ &= \text{FLOOR.MATH}(51/51,1)\\ &= \text{FLOOR.MATH}(1,1)\\ &= 1\end{aligned} \tag{5.76}$$

12. Complex Coordinate. The complex coordinate of each domain evaluation point is computed based on the following data:

(a) Each point’s column number and row number.

(b) The values of dx and dy.

(c) The length of the problem domain from the **Problem Data** tab.

***Excel* Implementation 5.3.3.9**

In this *Excel* application, the value of dx is stored in cell C14, and the value of dy is stored in cell C15. As before, let col_k denote the

k^{th} cell of the "Column Counter" column, and let row$_k$ denote the kth cell of the "Row Counter" column. Let L denote the length of the problem domain from the **Problem Data** tab. Let z_k denote the complex coordinate of the kth domain evaluation point. Then,

$$z_k = \text{COMPLEX(L}-\text{row}_k\text{*\$C\$14,col}_k\text{*\$C\$15)} \qquad (5.77)$$

13. Potential Value and Streamline Value. These are the CVBEM approximations of the target potential function and stream function at each of the domain evaluation points. These approximations are computed using the "Data Table" feature of *Excel*. The instructions are as follows.

Excel Implementation 5.3.3.10

For the potential value approximation:

(a) Identify the "Potential Approximation" cell, J3.
(b) In the blue cell of the "Potential Value" column, enter the formula =J3. The red cell must remain empty. Initially, all of the cells below the blue cell will be blank.
(c) Select the red cell, the blue cell, and all of the yellow cells (this will automatically select all of the empty cells below the blue cell that are adjacent to the yellow cells).
(d) In the "Data" tab of *Excel*, select "What-If Analysis".
(e) Select "Data Table...".
(f) Leave "Row input cell" blank.
(g) Enter $J$2 in "Column input cell".
(h) Click "OK". This will fill in the CVBEM approximations of the potential function at all of the domain evaluation points.

Excel Implementation 5.3.3.11

For the streamline value approximation:

(a) Identify the "Streamline Approximation" cell, X3.

(b) In the blue cell of the "Streamline Value" column, enter the formula =X3. The red cell must remain empty. Initially, all of the cells below the blue cell will be blank.

(c) Select the red cell, the blue cell, and all of the yellow cells (this will automatically select all of the empty cells below the blue cell that are adjacent to the yellow cells).

(d) In the "Data" tab of *Excel*, select "What-If Analysis".

(e) Select "Data Table...".

(f) Leave "Row input cell" blank.

(g) Enter $J$2 in "Column input cell".

(h) Click "OK". This will fill in the CVBEM approximations of the stream function at all of the domain evaluation points.

5.3.4 The Approximation tab

As discussed in Section 5.1.4, the **Approximation** tab consists of two matrices:

1. The first matrix consists of values obtained by evaluating the real part of the CVBEM approximation function at various points in the problem domain. These values are used to produce the contour plot of potential values in the **Plots** tab.

2. The second matrix consists of values obtained by evaluating the imaginary part of the CVBEM approximation function at the same points in the problem domain. These values are used to produce the contour plot of streamline values in the **Plots** tab.

The **Approximation Tab** is updated automatically and does not require user interaction.

5.3.5 The Plots tab

The **Plots** tab is used to visualize the computational results from the CVBEM model. Specifically, in Figure 5.11 at the beginning of this section, we have depicted the locations of the nodes and collocation points used in the CVBEM model. In Figure 5.18, we have depicted contours from the CVBEM approximation of the potential function. And in Figure 5.19, we have depicted contours from the CVBEM approximation of the stream function. These figures have been developed according to the instructions listed in Section 5.1.5.

Figure 5.18: Contours produced in *Excel* from the real part (potential function) of the CVBEM approximation function. These contours are from a CVBEM model of potential flow around a cylindrical obstacle.

Figure 5.19: Contours produced in *Excel* from the imaginary part (stream function) of the CVBEM approximation function. These contours are from a CVBEM model of potential flow around a cylindrical obstacle.

Bibliography

[1] Hromadka, T.V. & Guymon, G.L., Application of a boundary integral equation to prediction of freezing fronts in soils, *Cold Regions Science & Technology*, **6**, pp. 115–121, 1982.

[2] Hromadka, T.V., *The Complex Variable Boundary Element Method*, volume 9, Springer-Verlag, Berlin, Heidelberg, 1984.

[3] Hromadka, T.V. & Guymon, G.L., A complex variable boundary element method: development, *International Journal For Numerical Methods Engineering*, **20**(1), pp. 25–37, 1984.

[4] Hromadka, T.V., The complex variable boundary element method: Applications in determining approximative boundaries, *Engineering Analysis*, **1**(4), pp. 218–222, 1984.

[5] Hromadka, T.V. & Pardoen, G.C., Application of the CVBEM to non-uniform St. Venant torsion, *Computer Methods in Applied Mechanics and Engineering*, **53**(2), pp. 149–161, 1985.

[6] Hromadka, T.V., Approximation of slow-moving interface phase change problems using a generalized Fourier Series and the Complex Variable Boundary Element Method, *Numerical Heat Transfer*, **11**(3), pp. 341–351, 1987.

[7] Hromadka, T.V. & Lai, C., *The Complex Variable Boundary Element Method*, Springer-Verlag, New York, 1987.

[8] Hromadka, T. & Whitley, R., A new formulation for developing CVBEM approximation functions, *Engineering Analysis with Boundary Elements*, **18**(1), pp. 39–41, 1996.

[9] Hromadka, T.V. & Whitley, R.J., *Advances in the Complex Variable Boundary Element Method*, Springer, New York, 1998.

[10] Demoes, N.J., Bann, G.T., Wilkins, B.D., Hromadka, T.V. & Boucher, R., 35 years of advancements with the Complex Variable Boundary Element Method, *International Journal of Computational Methods and Experimental Measurements*, **7**(1), pp. 1–13, 2018.

[11] Hromadka II, T. & Guymon, G., Complex polynomial approximation of the Laplace equation, *ASCE Journal of Hydraulic Engineering*, **110**(3), pp. 329–339, 1984.

[12] Yen, C. & Hromadka II, T., Complex polynomial approximation of two-dimensional potential problems using generalized Fourier series, *Microsoftware for Engineers*, **2**(2), pp. 75–81, 1986.

[13] Hromadka, T.V. & Whitley, R.J., Approximating three-dimensional steady-state potential flow problems using two-dimensional complex polynomials, *Engineering Analysis with Boundary Elements*, **29**(2), pp. 190–194, 2005.

[14] Poler, A., Bohannon, A., Schowalter, S. & Hromadka II, T., Using the complex polynomial method with Mathematica to model problems involving the Laplace and Poisson equations, *Numerical Methods for Partial Differential Equations*, **25**(3), pp. 657–667, 2009.

[15] Bohannon, A. & Hromadka, T., The complex polynomial method with a least-squares fit to boundary conditions, *Engineering Analysis with Boundary Elements*, **33**(8–9), pp. 1100–1102, 2009.

[16] Haberman, R., *Applied Partial Differential Equations with Fourier Series and Boundary Value Problems*, Pearson, Upper Saddle River, NJ, fifth edition, 2013.

[17] Wilkins, B.D., Greenberg, J., Redmond, B., Baily, A., Flowerday, N., Kratch, A., Hromadka II, T.V., Boucher, R., McInvale, H.D. & Horton, S., An unsteady two-dimensional complex variable boundary element method, *SCIRP Applied Mathematics*, **8**(6), pp. 878–891, 2017.

[18] Wilkins, B.D., Flowerday, N., Kratch, A., Greenberg, J., Redmond, B., Baily, A., Hromadka, T.V., Hood, K., Boucher, R. & McInvale, H.D., A computational model of groundwater mound evolution using the Complex Variable Boundary Element Method and generalized Fourier series, *The Professional Geologist*, **54**(1), pp. 12–14, 2017.

[19] Wilkins, B.D., Hromadka, T.V. & Boucher, R., A conceptual numerical model of the wave equation using the Complex Variable Boundary Element Method, *Applied Mathematics*, **8**(5), p. 724, 2017.

[20] Hromadka, T.V., Linking the Complex Variable Boundary-Element Method to the analytic function method, *Numerical Heat Transfer*, **7**(2), pp. 235–240, 1984.

[21] Hunt, B. & Isaacs, L., Integral equation formulation for ground-water flow, *Journal of the Hydraulics Division*, **107**(10), pp. 1197–1209, 1981.

[22] Wilkins, B.D., Hromadka, T.V., Johnson, A.N., Boucher, R., McInvale, H.D. & Horton, S., Assessment of complex variable basis functions in the approximation of ideal fluid flow problems, *International Journal of Computational Methods and Experimental Measurements*, **7**(1), pp. 45–56, 2019.

[23] Barone, G., Pirrotta, A. & Santoro, R., Comparison among three boundary element methods for torsion problems: CPM, CVBEM, LEM, *Engineering Analysis with Boundary Elements*, **35**(7), pp. 895–907, 2011.

[24] Johnson, A.N., Hromadka, T.V., Hughes, M.T. & Horton, S.B., Modeling mixed boundary problems with the Complex Variable Boundary Element Method (CVBEM) using Matlab and Mathematica, *International Journal of Computational Methods and Experimental Measurements*, **3**(3), pp. 269–278, 2015.

[25] Hromadka, T.V., *A Multi-Dimensional Complex Variable Boundary Element Method*, volume 40 of *Topics in Engineering*, WIT Press, Southampton and Boston, 2002.

[26] Kirchhoff, R.H., *Potential Flows Computer Graphic Solutions*, CRC Press, New York, 1985.

[27] Johnson, A.N., Hromadka, T.V., Carroll, M., Hughes, M., Jones, L., Pappas, N., Thomasy, C., Horton, S., Whitley, R. & Johnson, M., A computational approach to determining CVBEM approximate boundaries, *Engineering Analysis with Boundary Elements*, **41**, pp. 83–89, 2014.

[28] Johnson, A.N. & Hromadka, T., Modeling mixed boundary conditions in a Hilbert space with the complex variable boundary element method (CVBEM), *MethodsX*, **2**, pp. 292–305, 2015.

[29] Wilkins, B.D., Hromadka II, T.V. & McInvale, J., Comparison of two algorithms for locating computational nodes in the Complex Variable Boundary Element Method (CVBEM), *International Journal of Computational Methods and Experimental Measurements*, **8**(4), pp. 289–315, 2020.

[30] Demoes, N.J., Bann, G.T., Wilkins, B.D., Grubaugh, K.E., Boucher, R. & Hromadka, T.V., Optimization algorithm for locating computational nodal points in the method of fundamental solutions to improve computational accuracy in Geosciences, *The Professional Geologist*, **56**(1), pp. 6–12, 2019.

List of figures

List of tables

Index

References followed by (t) denote a table and (f) denote a figure

www.ingramcontent.com/pod-product-compliance
Lightning Source LLC
LaVergne TN
LVHW011516200926
848484LV00001B/48
9781784664510